# The Compass Book

**Maintain**
**Repair**
**Adjust**
**Your Own Compass**

# The Compass Book

## Maintain, Repair and Adjust
## Your Own Compass

By Mike Harris

**Paradise Cay Publications**
Arcata, California
USA

Printed in the United States of America
Library of Congress Catalog Card #:
ISBN: 0-939837-27-7

First Printing 1998

Inquiries should be addressed to:

Paradise Cay Publications
P. O. Box 29
Arcata, CA 95518-0029
U. S. A.
800-736-4509 or 707-822-9063

2  3  4  5  6  7  8  9  0

# Acknowledgements

In writing the Compass Book, I am most grateful for the many useful suggestions and advice I've received from cruising boat crews in New Zealand and the Pacific. In particular my thanks to Mark Earl, Shane Johnston, Jim Johnston, Barry Young and Vernon Wall for reading and advising on the text, and to Russell Davis, Chris M Goulet. and Steve Loutrel for help with technical contents and software reviews. Also to staff at Electronic Navigation Ltd., and Trans Pacific Marine Ltd. of Auckland and to Paul Wagner of Autonav Marine Systems for their assistance and contributions.

Lastly, and by no means least my special thanks go again to my wife Di for her painstaking work on checking the text and preparing the illustrations.

# List of Illustrations

## Drawings

# Black & White Photographs

# Contents

# Terms and Conventions

| Term. | Meaning. |
|---|---|
| Agonic | Area where magnetic variation is nil. |
| Almanac | Table of predicted positions of astronomic objects. |
| Alnico No. 5 | Alloy used to manufacture compass magnets. 24% Co, 14% Ni, 8% Al, 3% Cu, 51% Fe |
| Altitude (In Astro or Celestial navigation) | The angular height of an object above the true horizon. |
| Azimuth | Angular distance of a point from the meridian. The azimuth of the sun or other astronomic object is its bearing measured from true north. |
| Azimuth Circle | A rotatable scale with sights that can be fitted to a compass for taking bearings of distant objects. |
| Bearing | The angular direction of one object from another or from a reference point. |
| Binnacle | A compass casing and support structure. |
| Bowditch | The book - "The American Practical Navigator" by Nathaniel Bowditch. |
| Cardinal Points | North, West, South and East compass points. |
| Coefficients (magnetic) | The 5 separrately identifiable components into which a complex deviating field may be resolved. |
| COG | Course Over the Ground. |
| Declination (magnetic) | See Variation. Declination is the preferred term in Earth Science. |
| Deviation | Deviation is a compass error caused by magnetism in a boat's hull, fastenings and equipment. It is a correction that's added to the compass heading to obtain Magnetic heading: Magnetic heading = Compass heading + Deviation |
| Dip | Angle between the horizontal and direction of the earth's magnetic field. |
| Dip Needle | A horizontally pivoted magnetic needle that's used to measure the vertical component of a magnetic field. |
| Directive force (H) | Horizontal component of the earth's magnetic field. |
| Diurnal | Daily. |
| External adjusters | Permanent or temporary magnets used to correct for deviation and fitted outside of the compass body. |
| Flinder's Bar | A soft iron corrector used to compensate for magnetism of the type caused by steamship funnels. Seldom used on small craft. |
| Fourier Analysis | Mathematical process by which a complex wave form can be resolved into simple sine and cosine components. |
| Gaussin error | Similar to Retentive error, though of shorter duration. Induced magnetism on a given heading may take up to two minutes to reach a stable value. |

| | |
|---|---|
| Gimbals | A pivotal fram used to hold an object in a horizontal position regardless of the angle of the structure to which it is attached. |
| H. O. | Hydrographic Office |
| Hard iron magnetism | See permanent magnetism. |
| Heading (magnetic) | The angle in degrees between the magntic meridion and the forward direction of the vessel's fore and aft line. |
| Heading (true) | The angle in degrees between true north and the forward direction of the vessel's fore and aft line. |
| Heeling Error | A compass error that occurs when the boat leans over. |
| Inclination | (synonymous with dip) |
| Induced magnetism | See Soft Iron. |
| Intercardinal Points | Northwest, Southwest, Southeast and Northeast compass points. |
| Internal Correctors | Correctors installed within the compass housing. |
| Isallogonic line | Line joining points of equal change of magnetic variation. |
| Isogonic line | Line joining points of equal magnetic variation. |
| Meridian (longitude type) | A plane that passes through the north and south poles and a point. |
| Naming conventions | Variation and Deviation are both positive quantities named EASTerly if measured in a clockwise direction. They are negative and WESTerly if measured in an anticlockwise direction. If variation or deviation lie between 180° and 360°, then subtracting 360° gives a more manageable figure and a change of name. |
| OSCAR QUEBEC | The international signal code used to identify a vessel that is swinging or correcting it's compass. |
| Pelorus | An instrument that uses a circular degree scale and rotatable sights, for measuring horizontal angles of distant objects. |
| Permanent Magnetism | Magnetism that is retained long after the magnetizing field is removed. |
| Quadrantal Correctors | One or a pair of adjustable, usually spherical, blocks of soft iron that are placed around a compass site to correct for errors that are a maximum on N, W, S and E, or NW, SW, SE and NE headings. |
| Quadrantal Error | Errors that are a maximum on N, W, S and E, or NW, SW, SE and NE headings. |
| Remalloy | Alloy used to manufacture permanent magnets. 12% Co, 17% Mo, 71% Fe |
| Retentive error | Caused by the tendency of a vessel to retain induced magnetism. A steel boat traveling in the same direction for several days under rough conditions may acquire magnetism from the earth's field that is held for a few hours after the course is changed. |
| Secular changes | Long period changes. |
| Soft Iron Magnetism | See temporary magnetism. |
| SOG | Speed Over the Ground. |

| | |
|---|---|
| Swinging the Compass | The process of checking the compass for deviation on a selection of headings. |
| Temporary Magnetism | Magnetism lasts only as long as the magnetizing field is present. |
| Total Error | The algebraic sum of Variation and Deviation. |
| Variation (Magnetic) | Angle between the magnetic meridian and the true meridian. It is a correction that's added to a magnetic heading to obtain the true heading:<br>True heading = Magnetic heading + Variation |
| Vertical Force Instrument | An instrument used in correcting Heeling error. |

# Preface

The magnetic compass is one of the most ancient of navigational instruments yet, even in these times of all electronic navigation, it remains a vital piece of gear for any small boat. Needing no electrical power, it works just as well under the water as above. In the harshest, roughest conditions its robust integrity makes it the one instrument likely to remain functional when all else is lost.

In spite of it's rugged construction, a good marine compass is also a sensitive instrument. For safe, reliable navigation it must respond to small changes in the relative direction of the earth's magnetic field but at the same time it must ignore any other source of magnetism from the boat on which it's installed. By tradition, the work of correcting for these effects and ensuring it's accuracy has been the exclusive domain of the professional adjuster and by the same tradition, many have tended to keep the 'secrets of the trade' very much to themselves. Unfortunately, as more and more commercial vessels have turned their reliance away from the magnetic compass, in many regions, the compass adjuster is becoming increasingly hard to find.

In writing the Compass Book, one of the main objectives has been to provide an introduction to the principles of compass work as a guide for those wanting to improve basic navigational skills or carry out as much of their own repair or maintenance work themselves as possible. It does not aim to seek to eliminate the need to for the professional adjuster but rather to explain the circumstances where the benefit of their training and experience would be of greatest advantage. The book is intended for owners of all types of small, non-commercial vessels, both power and sail, but should be of special interest to those traveling to remote parts where self reliance is a key to safety and survival.

# Chapter 1 - Introduction

From early school experiments with magnets it is easy to appreciate the forces that make a compass work. Just take a length of magnetised iron, hang it from a thread or support it from a pin and you have a simple compass. In basic form they are cheap to make and are produced by the million for everything from orienteering, hiking, car accessories or toys and games. For a marine compass, performance requirements are more demanding and money spent on a good instrument is returned in sound engineering, corrosion resistance, effective damping and, above all, a high level of consistent accuracy.

## Why is accuracy important?

If a steering or autopilot compass has a one degree error, after a day's run of run of say 60 miles, the boat's position will be in error by one mile. This relationship, one mile for each degree, is an approximation for small angles only though worth keeping in mind, particularly when deciding a safe clearance for a danger area or safe landfall.

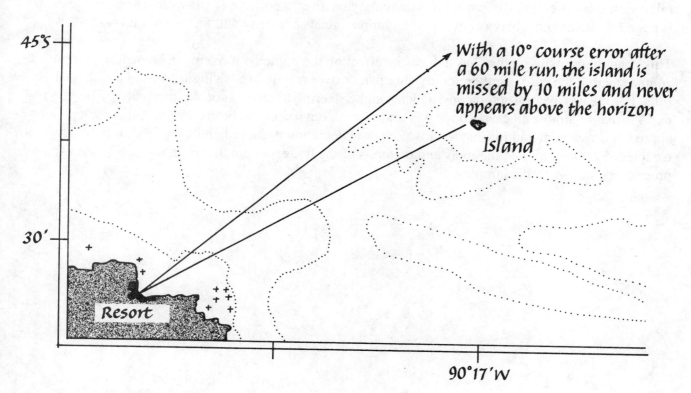

*Fig 1 Effects of a 10° course error.*

Under stable conditions, 95% of readings taken from a typical, well-adjusted marine compass are likely to be within 1° of the true value. Unfortunately, even in a moderate sea, the movement of a small boat can cause this error to increase by 5 to 10 times or even more as conditions worsen. So how is it possible for small boat sailors to make reliably accurate landfalls after distances of several hundred miles?

As the boat pitches and rolls, the compass card swings to either side of the true reading. By making a mental note of the center point of opposite swings and using this as a heading reference, an experienced helmsman is able to steer a steadier course. This is a technique that many people find instinctive, while others find it improves with practice.

While single spot readings can be very erratic, this method gives an average and more precise value though it can never improve upon the instrument's intrinsic accuracy.

Making sure the compass is well adjusted and being aware of its errors is an essential part of compass navigation - and also the main purpose of this book. We begin in Chapter 2 by looking at some basic principles; then in Chapter 3 we look at the constructional details of some common types of compass and their maintenance requirements.

## Compass Deviation - An insidious problem

Most navigators are aware of variation and deviation as corrections to be applied to the compass reading. These are described more fully in chapter 2. Deviation, caused by the magnetic effects of iron and steel parts of the boat present special difficulties. This is because it can change as the boat alters direction. With a well adjusted compass, for every degree that the boat turns, the compass should show a corresponding one degree change in heading. If deviation is present, this simple relationship is upset. On some headings there may be a large positive error while on others the error is negative, with points between where the compass reads correctly and the error is nil.

This changing sensitivity also has a dynamic effect on the compass movement. Depending on the course steered, the card is either unduly sluggish or over responsive, swinging wildly as the boat rolls. For the navigator, such inconsistencies make steering a reliable course impossible. The good news is that in almost all cases, once it is understood, deviation can be corrected. A methodical approach is essential and in Chapter 4 we see how deviation is caused and measured. Finally, in Chapters 5 & 6 we look at the tools and methods of the trade and put theory into practice with some procedures and worked examples.

# Chapter 2 - Basic Principles

*Marine Charts - 'The falsest thing in the world'*

Early Chinese were probably the first people to recognize that a piece of magnetised material suspended on a thread would always align itself in the same direction. In this case, it was a piece of iron-bearing rock or lodestone. In the late 13th century, a sample was shown to European navigators and mapmakers by the Italian explorer Marco Polo.

In studying this odd behavior more closely, it was later realized that the north/south alignment of the lodestone or compass didn't quite correspond to the established idea of the direction of true north. This had been determined from astronomic methods and a strong debate ensued to explain the difference or *magnetic variation* as it is still known. For a time this brought the whole validity of seagoing charts into question. Writing in 1542, Dr Pedro Nu◊ez notes that some skilled Pilots declared them "a mais falsa causa do mundo' (the falsest thing in the world)[1].

Mediterranean navigators noticed several difficulties when sailing east and returning on the reciprocal compass course. Ship captains crossing the Atlantic from Europe to Guadaloupe were advised to sail southwest until they reached 16° 20'N, using sun sights to determine latitude. At this point, says Alonzo de Santa Cruz in 1545, the compass needle 'northwests' a little and sailing due west by compass brings the ship to the rocks and banks off Trinidad (10° N to 11°N) and possible destruction. For many years, confusion reigned. So-called 'variation corrected' compasses were manufactured and some pilots believed that variation was due to a property of the particular lodestone used to magnetize or 'touch' the needle.

With the benefit of a few more centuries of further research and investment, today's ideas of magnetic variation are entirely different. Thanks to satellite technology, the latitude/longitude co-ordinates of any geographical feature are now known to within a few metres or less. However, the magnetic compass remains a top ranking navigational tool and charts showing the difference between true and magnetic north are still important references.

## How is the Earth's Magnetic Field Produced?

A simple question, but one that's not easily answered in simple terms. As a beginning, it's convenient to imagine the field to be produced by a giant subterranean bar magnet located inside the earth and running through it's center at an angle of about 15° to the rotation axis. In the northern hemisphere, one end would be buried beneath Ellef Ringnes Island in Canada (78.5°N 103.4°W) and in the southern hemisphere beneath Commonwealth Bay in Antarctica (65°S 139°W). Fig 2 shows the arrangement.

---

[1] Jean Rotz and the Marine Chart, 1542 E. G. R. Taylor. Journal of Navigation 1954 (Vol. 7 p. 138).

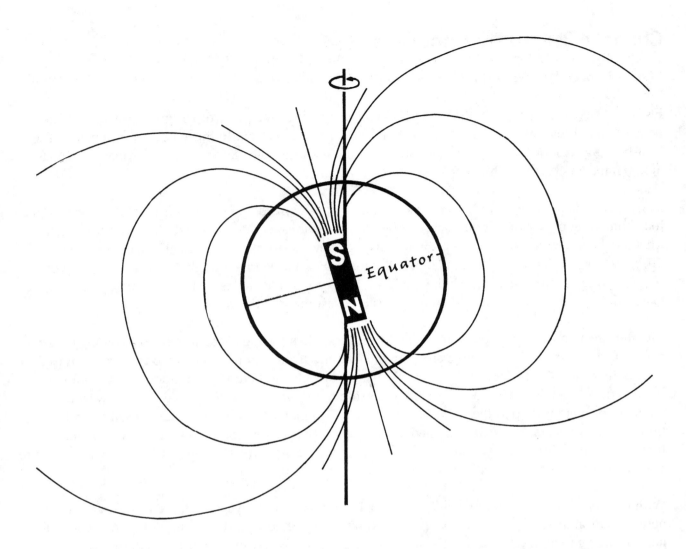

*Fig 2 **The Earth's magnetic field as produced by an imaginary bar magnet. Note that the magnet's north pole is in the southern hemisphere.***

For some purposes, the bar magnet theory is sufficient though Earth Science studies and years of Geomagnetic data have raised many difficulties. Instead of variation being constant, as one would expect of a bar magnet, long and short-term changes in strength and direction have been revealed. Short term or *transient* variations can be related to the position of the Sun (S type) or Moon (L type). These are distinguished from the long or *secular* variations that take place over many years and of which *Magnetic Variation* is of special interest to navigators. This is described in more detail on page 14.

The mechanism causing the Earth's field is not easily explained though it is thought to arise from the differential rates of rotation of the Earth's molten but conductive nickel/iron core and its more solid mantle. In this way, the planet behaves as a self-exciting dynamo, producing internal electric currents and an associated magnetic field. Further complicating the picture are the existence of thermal convection cells within the core. The flows of molten material circulating between the hot center core and cooler mantle act like wires producing more electric currents and magnetism. Study of these characteristics is known as *magnetohydrodynamics* and that of their change throughout the Earth's history is referred to as *paleomagnetism*.

By examining samples obtained by drilling through certain rock types, it is possible to construct a picture of how the field has changed over the ages. Surprisingly, it seems that the orientation of our present north and south poles has not always been as it is at present. Every 200, 000 to 300,000 years they switch direction. In terms of geological time, where the odd millenium is of little significance, these changes appear to have occurred quite quickly and a further reversal could be due shortly.

## Magnet Fundamentals

Most people have played with magnets at some stage in their lives, even if only with those toy 'bugs' that stick to the refrigerator. The basic idea; that if you put them round one way they stick together but if you put them round the other way they pull apart, is soon grasped and summed up in the old adage 'like poles repel - unlike poles attract'. So for a compass needle to be attracted to the Earth's magnetic pole in its northern hemisphere, it becomes evident that this pole must in fact be a 'South' magnetic pole. Hence the orientation of the Earth's Bar magnet in Fig 3.

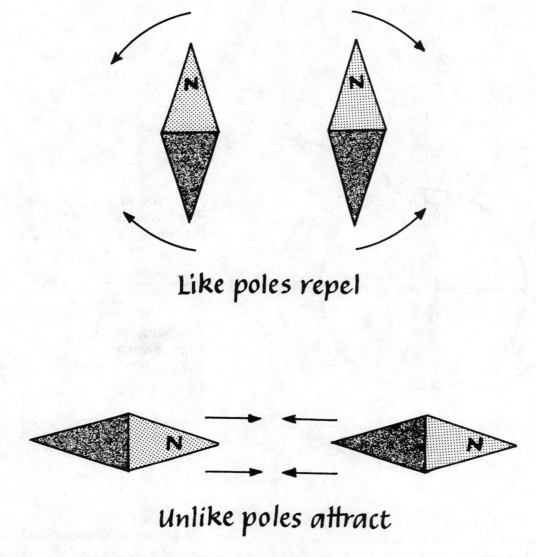

Fig 3 *How two magnetic needles behave when they are moved together*.

After experimenting in this way, it also becomes at least intuitively obvious that the forces of attraction and repulsion decrease as you move the magnets further apart. With these two concepts firmly in mind, you have mastered the basics needed to understand the principles of compass adjustment explained later in Chapters 5 & 6. Other than book-keeping skills, no deeper mathematical or physical theories are required as complex magnetic fields can be understood by breaking them into separate components. But before discussing these topics, for the remainder of this chapter there are some more terms and quantities that need to be explained.

### The Dip effect.
A compass needle aligns itself with the direction of the local magnetic field. A conventional vertical pivot allows the card to rotate freely in the horizontal plane but restricts vertical movement. However, magnetic fields are 3 dimensional and the vertical component can be significant. At the equator, the field direction is entirely horizontal while at the magnetic poles, the direction is entirely vertical. A practical effect of dip is that a compass card tends slope down towards the north in the northern hemisphere and the south in the southern hemisphere.

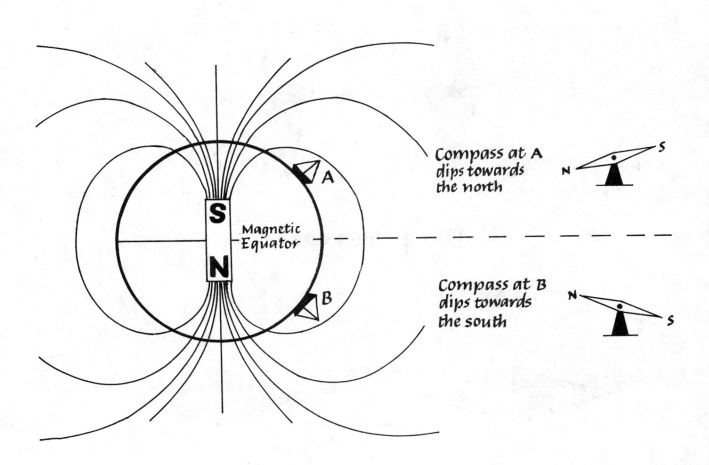

*Fig 4 A compass that's perfectly balanced at the equator tends to dip towards the north in the northern hemisphere and south in the southern hemisphere.*

## Magnetic Variation

Over most of the Earth, the north end of a compass needle points a little to the east or the west of true north. The angular amount of this difference is known as magnetic variation or simply variation and is usually printed on navigational charts along with its annual rate of change at the time of publication. Confusingly, Earth Scientists prefer to call it declination. This meaning of declination will not be used here as for navigators it is the name given to a coordinate in celestial/astro navigation.

Special charts with Isogonic lines (lines of constant variation) showing variations over a large area are available from the major hydrographic agencies (see page 59). Also available, are Isallogonic charts showing rates of change of variation though these are far less common. Predicting these rates is not an exact science so for critical applications it is preferable to use only the most recent data or charts that are less than 5 years old.

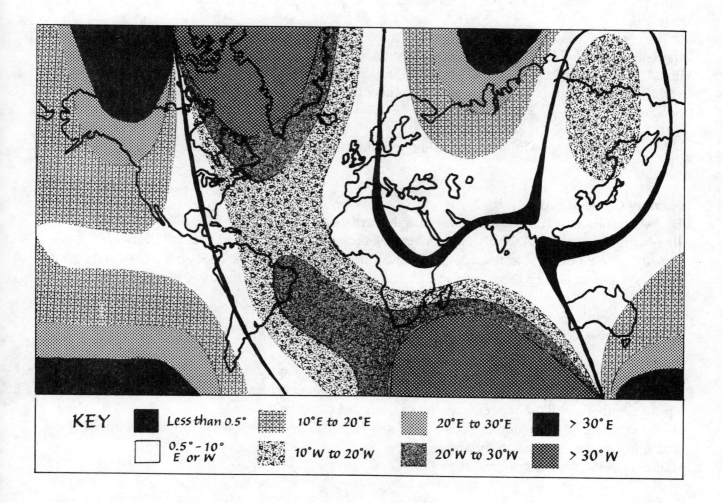

*Fig 5 Simplified chart of Magnetic Variation (Computer generated using Geomag)*

The earth's magnetic field undergoes daily and yearly changes. For our purposes, secular or long period changes are of most importance. For this reason, magnetic variations given on navigational charts are usually accompanied by an annual rate of change to allow future variations to be calculated. These rates can vary between around 2 and 25 years per degree. As an alternative to a chart you could use a computer program to obtain current variation and other magnetic data. Appendix 2 on page 53 describes some free software for this purpose. Many GPS receivers are also able to provide variation data though at the time of writing one major manufacturer was reported to be using data that was almost a decade old.

In summary, remember that variation depends upon your position on the earth and that it is a correction that's *added* to a magnetic compass heading in order to convert it to a true heading.

**True Heading = Magnetic Heading + Magnetic Variation**
or by rearrangement:
**Magnetic Heading = True Heading - Magnetic Variation**

## Compass Deviation.

Deviation is a compass error caused by magnetism in a boat's hull, fastenings and equipment. These parts produce fields that interfere with the compass, causing it to deflect from the direction of magnetic north.

Remember (from page 14 ) that, unlike variation, deviation is a local error that may change depending on which direction the vessel is heading and can seriously affect the compass sensitivity. As with variation, it is a correction that's *added* to an uncorrected compass heading to obtain the magnetic heading.

**Magnetic Heading  = Compass Heading + Deviation.**

### Naming Conventions

By long tradition, variations and deviations that are measured in a clockwise direction, to the east of true North are named **EAST**erly and conversely, those that are measured in a anti-clockwise direction, to the West of true North are named **WEST**erly. In handling these quantities in calculations, it's more convenient to treat them as positive or negative numbers using the usual algebraic rules:

*Clockwise angles (ie Easterly angles) are Positive (+) quantities*
*Anti-clockwise angles (ie Westerly angles) are Negative (-) quantities*

Similarly, latitudes and longitudes used in calculations are also assigned as positive or negative quantities following the usual graph coordinate conventions; ie measurements made to the right or up are positive and those made to the left or down are negative.
So for example, in calculations a position of 46° 15'N 028° 30'W, by converting to decimal degrees and applying the signs would become:-
Latitude: +46.25°, Longitude: -28.5°

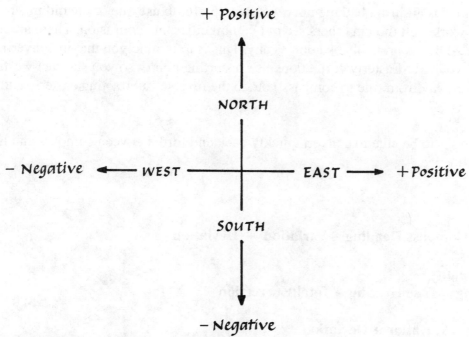

Fig 6 Cartesian coordinate conventions

## Applying Variation and Deviation Correctly

Adding or subtracting Variation and Deviation to convert between compass readings and true north related bearings is almost a trivial procedure though in times of stress one that frequently causes difficulties. To help visualize the process, some charts are printed with two concentric compass roses; the outer orientated to true north and the inner one to magnetic north. No matter in which direction you need to convert, the process is basically the same. Fig 7 shows the details. Using this same principle you may consider making a more permanent device for handling conversions. Photocopy a pair of compass roses onto stiff card. Cut them out and with a pin through their centers you have a useful tool for applying deviation and correction.

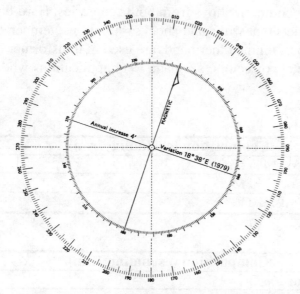

*Fig 7 **Dual True and Magnetic compass roses printed on a chart. In this case for an area where variation was 18°38'E (1997). To convert between magnetic and true (or vice versa) simply lay a ruler between the center and the value you are converting from. Read the converted value from the adjacent scale.***

If high school math was all just a bad memory, you may prefer to use one of the old mnemonics. For some people they work well though others see only ambiguities and confusion. There are heaps to choose from but CADET is probably as good as any. This is to remind you that in converting from **C**ompass to **T**rue, you **A**dd **E**asterly corrections. It's a starting point - so you subtract west, and in converting the other way from true to compass you do the reverse, subtracting easterly and adding westerly corrections.

The important thing is to be able to convert quickly back and forth between compass and true values with absolute confidence.

**To Summarize:**

**True Heading = Compass Heading + Variation + Deviation**

So by rearrangement:
**Compass Heading = True Heading - Total Correction**
Where:
**Total Correction = Variation + Deviation**

# Types of Magnetism

Iron is the material responsible for causing most compass deviation and is the main ingredient of all types of steel. These and other iron alloys can be broadly divided by their magnetic properties into two types - hard and soft. Hard materials are able to retain their magnetism over a long period of time (*permanent* magnetism), whereas in Soft materials, it is held only as long as the magnetizing field is maintained (*temporary* magnetism). For example, if a thin rod of soft iron is aligned in a magnetic North to south direction, it acquires magnetism from the earth's field but of opposite polarity ie it's north facing end becomes a north pole and south facing end a south pole. If the same rod is now turned in a magnetic East West direction, all magnetism is lost.

The following table lists some common alloys from both categories. Hard types are used principally in applications requiring magnets that withstand shocks, vibrations, temperature changes, ageing and demagnetizing fields. Soft types, on the other hand, are used in transformer laminations, for magnetic shielding and in quadrantal correctors (see page 39) - applications where an ability to remagnetize is important.

## Hard Magnetic Materials

| Alloy | Composition (excluding Iron) |
|---|---|
| Alnico No. 5 | 24% Co; 14% Ni; 8% Al; 3% Cu |
| Remalloy | 12% Co; 17% Mo |
| Tungsten steel | 6% W; 0.7% C |

## Soft Magnetic Materials

| Alloy | Composition (excluding Iron) |
|---|---|
| Mu-metal | 5% Cu; 2% Cr; <1% Mn |
| Permalloy | 78% Ni |
| Upermalloy | 79% Ni; 5% Mo; 0.5% Mn |

**What types of objects are likely to become temporary or permanent magnets?**
Cast iron parts, such as engine blocks, motor frames etc. are predominantly soft iron but mild steel, the basic building material of all steel vessels, has characteristics of both types. When a steel vessel is built, heavy welding currents, hammering and heating cause local changes in the material's magnetic properties. Over the following weeks these may change, rapidly at first then becoming more stable as the months pass. Also, if major repairs are carried out to an older vessel, its magnetic properties are likely to be affected.

**Is compass deviation only a problem for steel boats?**
No. Large amounts of deviation can occur even on glass fibre or wooden boats. The engine, machinery, wiring, electrical equipment, onboard gear, bicycles, tinned food, or anything that has iron in its construction can produce deviation. Taken as a whole, the pattern of permanent and temporary fields can be extraordinarily complex. Plotting a graph of deviation as the boat's heading is moved through a full 360° can produce a complex curve of the type shown in Fig 30a on page 41.

Through a mathematical method known as Fourier Analysis, it is possible to break down this type of curve into simple sine and cosine components. The technique is widely used throughout physics in applications from acoustics and radio communications to tide and current studies. In some of these, there may be hundreds of separate constituents to consider and the mathematics can seem lengthy and obscure. Fortunately, in compass work, any deviation pattern can be described in terms of just 5 separate components and simple solutions are possible. These are further explained in Chapter 5.

# Chapter 3 - Compass Types - Installation and Maintenance

Compasses are made in a wide range of types and styles and to a large extent choice is a matter of personal preference. A useful question to ask is if internal repairs are possible.

These days, many are built as 'sealed for life units' and replacing seals, re-balancing the card or exchanging worn pivots may be difficult even for a well equipped instrument workshop. The idea being that when such jobs are necessary, you simply buy a new unit.

Unfortunately, for long distance cruisers, good compass stores are not always easy to come by and the ability to carry out patch repairs with tools already on board is a useful attribute even if the results are less than perfect and carry no warranty.

Except for considerations of weight and space, on the whole, the bigger and more readable the scale, the better, though this still leaves a bewilderingly large choice. In comparing different types, it's convenient to divide them according to the type of gimbals used.

## Externally Gimbaled Compasses

These are the traditional type, still used today in older vessels and some life boats. Prior to the 1970s these were very popular and from their rugged, simple design it's easy to pick out the main working parts. Fig 8 shows typical constructional details.

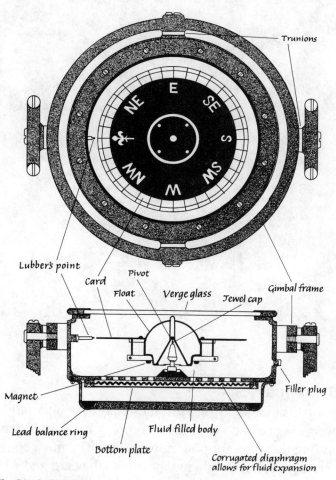

*Fig 8 **A Traditional Externally Gimbaled Compass***

**The grid compass.**

This is a particular type of externally gimbaled compass and a personal favorite. A few are still manufactured though they are not the most popular. The main feature is a rotatable bezel surrounding the main card. Around the outside is an additional degree scale that can be lined up against the lubber line and running between 0° and 180°, across the diameter, are a pair of engraved lines that are lined up against a similar pair of lines on the card below.

An advantage of this arrangement is that if the bezel is set to the required course, all the helmsperson has to do is keep the lines on the card and lines on the bezel together. After a long, tiring stint at the wheel this is often a lot easier than watching numbers fly by and having to remember the correct one to steer to. At night, since it's not necessary to actually read the numbers from the scale, illumination can be minimal. Just a couple of luminous markings on the card and bezel are often all that's needed.

*Photo 1 A Grid Compass*

## Internally Gimbaled Compasses

Most compasses fitted to small sail and power craft these days are internally gimbaled. In these instruments, the gimbal frame is enclosed within the compass body and immersed in the same fluid as the card, the fluid providing damping for both movements. The most conspicuous feature of these compasses is the domed, transparent hemisphere that covers the card and which is necessary to provide fluid space for the gimbal frame to swing past.

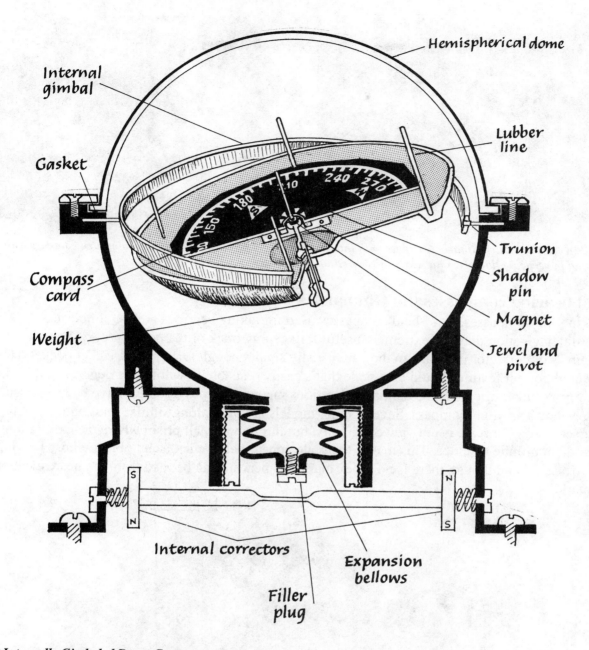

Hemispherical dome

Internal gimbal

Lubber line

Gasket

Compass card

Weight

Trunion

Shadow pin

Magnet

Jewel and pivot

Internal correctors

Expansion bellows

Filler plug

*Fig 9 Internally Gimbaled Dome Compass*

## Variations on a theme

Compasses of this type are often mounted on a binnacle and enclosed within a cylindrical aluminum or plastic case. As an alternative, they may be made for mounting through a hole in a horizontal fascia, with only the card and hemisphere visible above the panel. Often they share space with other instruments and controls which if too close may be a cause of deviation.

Vertical bulkhead mounted compasses are also popular, particularly on smaller tiller- steered sailing boats where there may not be room for a binnacle. Often they are mounted through a cockpit bulk-head to the side of the main hatch. With this type, the scale is generally marked on a vertical band attached to the edge of the card as in Photo 2 b. This is sometimes called a *Combination card* compass as it also includes a scale on the top of the card that can be read from above.

25

*Photo 2 (a & b) a)  Saura Binacle Compass with internal correctors b) Plastimo bulkhead mounted compass with combination vertical and horizontal scales, also a clinometer scale.*

## Hand bearing compasses (no Gimbals)

Of all marine compasses, hand-held compasses used for taking bearings are the simplest.

In coastal navigation they are extremely useful for keeping track of the boat's progress and having no gimbals or corrector mechanism they are usually small enough to carry in a jacket pocket. However, an important feature should be a scale that's easily readable to within one degree. At this size some form of magnification is essential and various ingenious types of optics have been devised. Photo 3 shows a 'Mini' compass. Several other models are built along similar lines with a small immersed card. The scale is not viewed directly, but through a small prism where the calibrations appear at an infinite distance. This makes the scale appear below the distant object whose bearing is to be measured and though there is no index mark, it's bearing can be read from the nearest calibration mark.

*Photo 3 The 'Mini' hand bearing compass in action.*

**Caution**

Hand bearing compasses have no means of correcting for deviation so in use care should always be taken to make absolutely sure that all possible sources of local magnetism are at least a couple of metres away. Aboard glass-fibre or wooden vessels with non ferrous fastenings, it is usually possible to find a spot where reliable bearings can be taken. With steel or ferro-cement boats the problems is more difficult. As a simple check, pick a distant land based object whose magnetic bearing can be determined from a chart. As the boat is turned slowly through a full 360°, if the site is deviation free, this should always correspond to its compass bearing.

In many cases, a convenient site may simply not exist. As a solution, it was once a common practice to fit an azimuth ring or bearing circle to a fixed compass that had been corrected for deviation. This device included sights for viewing distant objects and by viewing across the compass card their bearings could be measured. The compass used was generally a binnacle mounted main steering compass though these days yacht designs have changed and this position rarely gives a full 360° unimpaired view of the horizon.

## Flux gate and electronic compasses

Flux gate compasses have been in existence for many years and were among the first type all electronic compass. These may contain no moving parts and instead of a pivoted needle, use a coil to sense the earth's magnetic field. In original form they were totally analog devices though these days much of the signal processing is carried out by digital techniques.

Main advantages are:

- A single sensor can drive several small 'repeater' displays in other parts of the boat and can be used to provide heading information for an autopilot, computer logging and control.
- Corrections for most types of deviation can be carried out under software control.
- The need to provides mechanical gimbals is in many cases eliminated
- The sensor can be placed almost anywhere on the boat

This last feature is especially useful if the display unit is to be placed close to other instruments or sources of magnetism. With a wider choice of locations, it's easier to find a spot where deviation is minimal though this does not mean that the sensor can be hidden and forgotten. The possibility of someone storing tinned food or gear with magnetic parts close by should not be ignored as the effect on navigation could be unexpected and serious.

Often a good deviation free spot can be found on the mast. Due to the extreme pitching and rolling movements, the masthead is not ideal but a position a couple of metres above the deck is a good compromise. A disadvantage to be considered is that should the boat be dismasted compass control would be lost. A general disadvantage of all electronic compasses is that they require power to operate. To guard against breakdowns, an emergency backup, perhaps a conventional compass, is essential.

In comparing flux gate to conventional compasses in autopilots, Paul Wagner (manager of Autonav Marine Systems writes:

*These direct sensing compasses are frequently claimed to be far superior to "old fashioned" fluid filled compasses. In fact, electronic compasses, have been in use for over 70 years and their limitations are well known to compass experts. They are in common use today mainly because they are less expensive to manufacture than the conventional fluid filled compass with its floated card, magnets, pivots, jewels and sealing system. The flux gate consists of a field sensor, usually an inductor, mounted to a gimbaled platform which is intended to sense the horizontal component of the earth's magnetic field. The earth's field has two components, the horizontal field which gives directional information and the vertical field which provides no useful heading information. If the sensor should move from its intended horizontal position due to roll, pitch or slamming in a seaway, the sensor will pick up some of the vertical field, mixing it with the horizontal field causing an error in the actual course. The same problem would occur in a conventional fluid filled compass except that the pivot and jewel offers a second line of defense in decoupling the sensor (card and magnets) from the vessel motion. It is a seldom recognized fact that this extra isolation from vessel motion, coupled with fluid damping, results in a conventional fluid compass having much greater stability than any electronic compass under most conditions.*

## A new generation of heading sensors
Driven by the need to provide reliable heading information, in industrial and military robot and vehicle applications, much development effort has been devoted to producing better sensors.

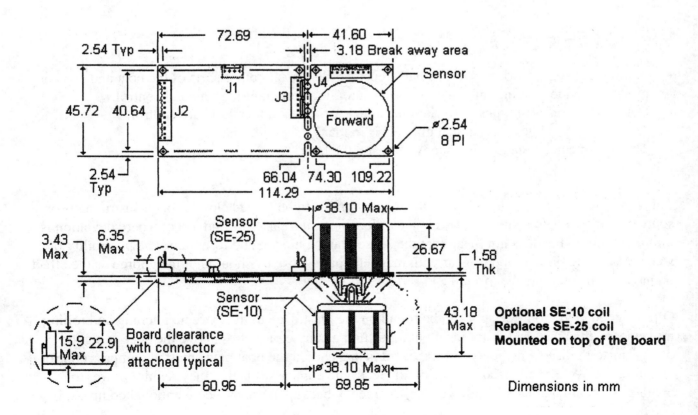

*Fig 10 Layout of the KVH Industries C-100*

The KVH Industries C-100 flux gate heading sensor (shown in Fig 10) is a basic 'compass engine' used in a number of marine instruments and autopilots from other manufacturers. It has an internally gimbaled solid state subsystem with a free floating ring core that keeps the sensor horizontal with respect to the earth. A current is driven through the ring core while coils at right angles to it sense the earth's field. Measurements are taken at second intervals and an onboard microprocessor applies filtering and averaging algorithms. Its technical specification is as follows:

| | |
|---|---|
| *Accuracy: | $\pm 0.5°$ or $\pm 10$ mils RMS (SE-25 coil assembly and digital outputs) |
| Repeatability: | $\pm 0.2°$ or $\pm 5$ mils (SE-25 coil assembly and digital outputs) |
| Resolution: | $0.1°$ or 1 mil |
| *Dip angle: | $\pm 80°$ (Maintains stated accuracy after auto-compensation up to $\pm 80°$ magnetic dip angle) |
| Tilt Angle: | $\pm 16°$ (SE-25 Coil)    Deviation = $\pm 0.3°$ RMS<br>$\pm 45°$ (SE-10 Coil - gimballed)  Deviation = $\pm 0.5°$ RMS |
| Input Voltage: | +8 to +18VDC or +18 to +28VDC (user selectable) |
| Current Drain: | Current Drain: 40 mA DC; maximum |
| Variation: | $\pm 180.0$ degrees adjustment range (offset) - user selectable |
| Index Error: | $\pm 180.0$ degrees adjustment range (offset) - user selectable |
| Time Constant: | 0.1 to 24 seconds - user selectable |
| Weight: | 64 grams (2.25 ounces) with SE-25 coil assembly |
| Reliability: | Mean Time Between failures greater than 30,000 hours |

As an improvement on the basic flux gate sensor, a recent development is the addition of dynamic rate gyros that measure the pitch, roll and azimuth of the vessel. Though this is still essentially a magnetic compass, compared with conventional mechanical gyros, it is roughly a tenth of the size, weight and cost. Output from the dynamic rate gyro is used to help correct the main heading data for the effects of pitch and roll as described by Paul Wagner.

A fuller description of the Digital Gyro, Fiber Optic Gyro and other related developments, can be found at the KVH Industries web site (see page 61).

**Electronic compass corrections**
Software correction of deviation is a standard feature in all currently produced heading sensors. Procedures vary, though for the C-100 sensor, three options are possible:

*Eight Point Auto-Compensation* - the host platform starts on an arbitrary heading and proceeds in a controlled rotation, stopping briefly at approximately 45° intervals to collect data. No specific heading is required. This method is commonly used with handheld or land-based instruments or vehicles which can easily turn and pause at eight different headings.

---

* Accuracy measurements apply to a level compass module after compensation. After installation and auto-compensation, typical accuracies of $\pm 0.5°$ are achievable on most platforms.

*Circular Auto-Compensation* - the host platform starts on an arbitrary heading, then proceeds through a slow (approximately 2 minutes) continuous 360° turn. No specific heading is required. This method of compensation is commonly used in vessels and wheeled vehicles where it is easier to steer through a gentle circle rather than stop at 8 points.

*Three Point Auto-Compensation* -the host platform starts at an arbitrary heading, then points briefly at two additional known headings approximately 120° apart. This method is used when known headings are available from an installed inertial reference or when a surveyed compass rose is available.

In some cases the software correction is unable to correct fully for all forms of deviation present (eg heeling error - see page 40). After completing corrections, it is always advisable to check for deviation by carrying out a conventional swing. If it persists, the traditional method of analysis and correction (explained in Chapter 6 - Compass Adjustments) can be used to remove it.

## Choosing a Site for a Conventional Fixed Compass

Since a steering compass may be used by the same person for many hours, it seems an obvious requirement that it should be placed somewhere where it's easy to read. If the same person is also responsible for keeping watch, then to avoid strain only small eye movements should be needed to switch between looking at the compass and the horizon ahead. Ideally, to achieve a reasonable compromise with other equipment, the compass site's special requirements should be considered at an early stage of construction. Unfortunately, this is often not the case and, particularly among commercially built power craft, the compass has to be fitted between CD players, loud speakers, drinks coolers and other creature comforts.

To minimize deviation, choose if possible a position on the centerline of the boat that is not surrounded by steelwork or close to any source of magnetic interference. Avoid placing it near single runs of high current cable, winch or radio power cables, window wipers, steel hinges or steel reinforced moldings. The list is by no means exhaustive but see page 31 for examples of other possible sources of deviation. As a rough guide, try to keep the compass at least one metre away from items of this type.

### Compass illumination

A box for a candle or kerosene lamp fitted to the side of the binnacle was once the norm but fortunately those days are gone. The small bulb of an electric compass light takes precious little current but to avoid deviation it's a good idea to make sure that the supply leads are twisted together. To maintain your night vision the light needs to be red and just bright enough to avoid eyestrain. Surprisingly little light is needed and a red LED (Light Emitting Diode) with suitable series resistor can work very well. LEDs are cheap, robust and durable.

Non electric alternatives include luminous paint or beta lighting. Luminous paint has been around for many years and has been used extensively on clocks and watches though for health reasons the practice is largely discontinued. The paint contains a small amount of radium that emits gamma rays and short range alpha particles. Light is emitted when the particles strike a fluorescent component that's combined with the paint. Light levels are fairly low and insufficient for reading numbers, but if it is simply needed to show the alignment of two marks, as on a grid compass, it is sufficient (see Photo 1 **A Grid Compass**. The illumination deteriorates over a period of several hours but can be 'recharged' by giving it a few minutes exposure to a bright light.

Beta lighting, frequently used in hand bearing compasses, consists of a small sealed cell containing a small amount tritium. This is an isotope of hydrogen that emits β particles that are again used to stimulate a fluorescent material giving a greenish light that's sufficient to read the compass scale. The isotope has a half life of 12.3 years which is about the useful life of the cell.

**Compensating for Dip (Inclination) of the earth's magnetic field.**
The dip effect is noticeable when a compass is moved to a latitude for which it was not balanced. As an example, the main steering compass aboard my own boat was manufactured for use in European waters and when traveling south to the Canaries and west across the Atlantic, it gave no problems. In the Southern Caribbean the card's south end developed a noticeable dip though not enough to be of concern. Moving on into the Pacific, across the equator and through French Polynesia, the dip increased greatly, but still the compass remained functional. Crossing the tropic of Capricorn (23°S) proved to be the limit, as at this stage the steeply inclined card would occasionally jam against it's mountings giving erratic readings. In the relative tranquility of North Minerva reef, I drained the oil and stripped the unit. The balance weight consisted of a small brass washer, attached with epoxy beneath the south side of the card. With this removed and the compass reassembled, normal balance was restored, and the reassembled compass again worked perfectly.

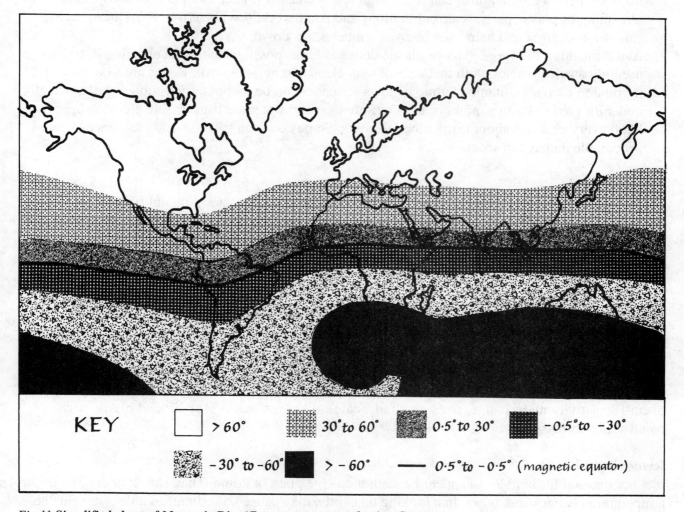

*Fig 11 Simplified chart of Magnetic Dip. (Computer generated using Geomag)*

In general, compasses with internal gimbals tend to suffer less from dip and by designing the card to have a center of gravity that lies below the pivot, manufacturers can reduce the problem still further. Several manufacturers produce different models or issue separate cards for particular regions. Fig 11 gives values of the dip component throughout the world. From this it can be seen that the range of values over most of the Northern hemisphere is relatively small and a card that is balanced for mid latitudes is usually good for the whole region.

In the Southern hemisphere, dip values change more quickly and travelers can expect to need more frequent compass changes. For circumnavigators using the popular trade wind routes (Panama, Torres Straits, Red Sea, North Atlantic) a compass balanced for the magnetic equator is a good compromise for the entire trip.

## Maintenance

A good quality marine compass is among the most enduring of nautical instruments. Precious little routine maintenance is required, though considerate treatment can do much to ensure a trouble-free life. In particular:

1.  Avoid high temperatures. Internal parts are often black and in direct sunlight, particularly in the tropics, internal temperatures can reach high levels causing higher internal pressures and premature aging of plastic parts, scale calibrations and paint-work. A white cover, even a canvas bag, can block sunlight and help keep internal temperatures down.
2.  Avoid touching the verge glass or plastic dome as far as possible and when cleaning it, be sure to use only a soft tissue or cloth that's absolutely clean and non-abrasive. Better not to wipe it at all if you don't have a suitable cloth to hand as scratches can be particularly troublesome, especially under dim red lighting at night or in rain. Soft soap and water are usually safe cleaning agents but be very cautious about using other cleaners. Some common solvents and cleaners can cause irreparable damage in seconds.

### Compass fluids

The main purposes of the compass fluid is to provide damping, buoyancy and lubrication for the card pivot and gimbals. In domed compasses, the fluid's optical properties are also important in providing magnification for the card's scale. Changes in ambient temperature cause the fluid to expand or contract and to keep internal pressures within safe limits, the bowl is fitted with a flexible diaphragm or bellows.

Compass fluids have to be non-corrosive and able to operate over a wide range of temperatures without significant changes in viscosity or refractive index. Often they are fine hydrocarbon oils or mixtures of ethanol and water (typically 45% ethanol; 55% water). When replacing or toping up the fluid it is most important to ensure that the type you are using is compatible with the fluid already in use. If the wrong type is used, there is a risk that the mixture may become cloudy and opaque, the overall sensitivity may change, seals may fail, scale calibration marks may lift and plastic parts could soften or dissolve.

### Removing a bubble

It is not unusual for bubbles to appear beneath the verge glass or dome. They shrink or enlarge as the temperature changes and, other than looking unsightly, may cause no difficulties unless they are so large that the normal movement of the card is affected.

A possible cause is that when filled, the fluid contained suspended air; a bubble forming as air pockets float upwards and combine. If the bubble is accompanied by signs of leakage the cause is more likely to be due to a crack or perished seal through which air is pulled as the pressure inside responds to a drops in ambient temperature. In this case, clearly the fault must be repaired before a long-term solution can be achieved.

Given the choice, a complete strip and re-assembly is a job best carried out by an instrument workshop specializing in compass repairs. Cleanliness is of primary importance particularly with compass fluids and difficult to achieve if you are not experienced at handling it. All containers must be kept scrupulously clean, as any dirt or drifting debris is likely to show up in the fluid after the compass is re-assembled. With internally gimbaled compasses, the dome's magnification makes dirt especially noticeable. Never shake the fluid and, before using, allow it to stand for a day to encourage air to rise.

Provided the compass has a filler plug, topping up to remove a small bubble is not a difficult job. After carefully manipulating the compass so that the filler is uppermost and with the bubble beneath, secure the compass in this position and remove the plug. Use a glass pipette to add an excess of fluid and re-screw the plug into position with a new sealing washer if necessary.

## Spares to carry
For long distance cruising, it's useful to carry a few compass spares. Even if you are not prepared to use them yourself, a shoreside workshop will be more able to assist if you can provide them with the parts needed. Items to consider include:

- Spare cards balanced for different regions.
- Top up fluid - sufficient for at least one complete replacement.
- Replacement dome or verge glass. This is a vulnerable part, sometimes smashed by accidental collisions.
- Spare Gaskets and seals
- Spare expansion bellows or diaphragm.
- Spare pivots - seldom needed.

# Chapter 4 - Swinging the Compass

Deviation, the distortion of the magnetic field by neighboring magnetic objects, can destroy the accuracy of even the best marine compass. Its effects were outlined in Chapter 1, but how do you know if it is present and how is it measured? Here we look at practical details of compass swinging and how it provides the answers.

The term 'Swinging the Compass' is a little misleading since it is the boat to which it is attached that is swung. The process involves turning the boat through a full 360° circle. If deviation is absent, the compass card will at all times remain pointing along the magnetic meridian. To measure deviation on a particular heading, we record the amount by which it has been deflected from this alignment, by taking a distant object, whose correct magnetic bearing can be independently determined from a chart or some other method, and comparing this with it's bearing as indicated by the compass. By repeating this process for headings at, say, 10° intervals we construct a curve of the type shown in Fig 30a

A good spot for the operation is a clear, open stretch of water with few other boats around and little current. If you can, pick a day with little wind and visibility clear enough to see your chosen reference object with ease. If you are using the sun, best times to choose are shortly after sunrise or before sunset, rather than the middle of the day when it's too high to measure accurately and its bearing is changing quickly.

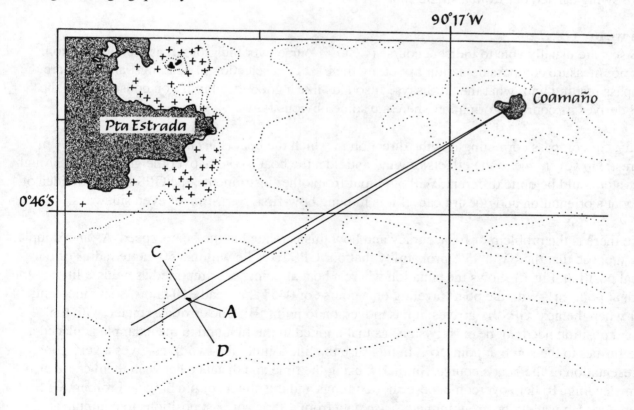

*Fig 12 **The more distant the reference object, the more room to swing the boat. To keep the Coama'o light bearing within 1° a boat at point A can move no more than 65 metres along the line C - D.***
***The following table gives the limiting distance for other ranges.***

## Choice of Reference Object

A suitable reference object could be a tower, lighthouse or mountain peak that's easily visible, clearly marked on the chart and more than two miles away. The greater the distance the better as this allows the boat more freedom of movement during the swing.

In the example shown in Fig 12, a swing is carried out at point **A** which is 2 miles from the reference point at Coama'o light. At this distance, the vessel can move no more than 65 metres along the line **C - D** if the bearing of Coama'o is not to change by more than 1°.

| Range (nautical miles) | 1 | 2 | 3 | 4 | 5 | 6 | 7 | 8 | 9 | 10 |
|---|---|---|---|---|---|---|---|---|---|---|
| Travel (metres) to change the bearing by 1° | 32 | 65 | 97 | 129 | 162 | 194 | 226 | 259 | 291 | 324 |

Sometimes it's difficult to pick a single object that can be clearly observed from the compass site for the whole 360° as the boat is turned. Perhaps a spray hood or window frame is in the way. To cover obscured sectors, a second or third object may be used and in Fig 12 the light on Pta. Estrada is a possibility though, since it is little more than a mile away, the boat would have little room to maneuver during the swing. Faced with these circumstances you may consider setting a series of anchors around the boat. By using winches to haul in or let out lines, the boat can be slowly rotated and the entire swing carried out from a single spot.

### Can we use GPS?

GPS sets are usually able to indicate course (COG - Course over the ground) as well as position; why not measure compass deviation by setting the boat on a selection of headings, and compare the compass reading with what the GPS says? It sounds like a good idea but if you consider how the GPS derives its course information, there are some obvious flaws.

Firstly, for compass correcting it is the direction in which the boat is *heading* (pointing) that is of interest. However, due to the effects of wind and tide, the boat's *course* over the ground and through the water could be quite different. Without input from other instruments, the GPS set has no idea of the boat's orientation and doesn't care if it is moving bow-first, stern-first or even sideways.

Next, there is the problem of fix accuracy and how this is related to the boat's speed. As an example, let's suppose that there is a 95% probability that our GPS fixes lie within a 50 metre radius of our actual position. Fig 13 shows the situation where a boat at point **'A'** is reported as being a little to the north at point **'a'**. With the boat traveling on a course of 045° at a steady 10 knots, some moments later when the next GPS fix arrives, it has moved on to point **'B'**. The error associated with the second position need not be exactly same as that applied to the first and in the example is placed a little further to the south at point **'b'**. In this case the line joining the two fixes gives a very poor representation of the boat's course. Similarly, using the time and distance between them could also be misleading. To help overcome these shortcomings and the effects of pitching and rolling, GPS computers can usually be programmed to use data from a series of past positions to compute an average speed and distance. For passage making it is an effective solution but for compass work on slow vessels, significant short-term course changes are not detected soon enough or may be missed altogether.

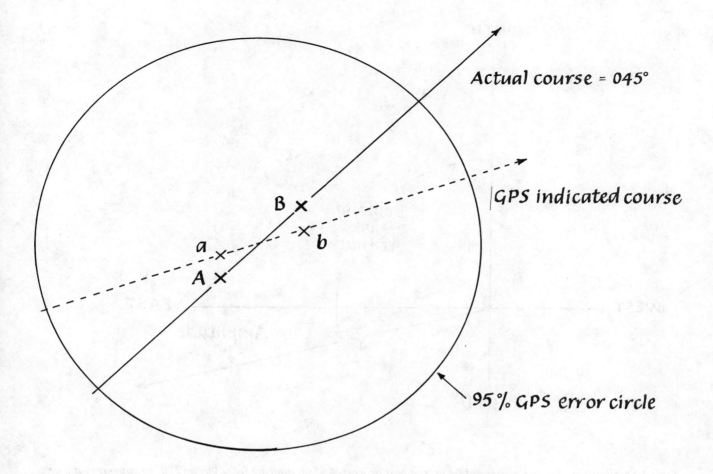

*Fig 13 At low speeds, GPS course indications can be misleading. Actual positions 'A' and 'B' could be reported as positions 'a' and 'b' and still lie within the 95% error radius, though the line joining them does not represent the true course.*

A more useful application for GPS in compass work is in fixing the boat's position on the chart so that bearings to reference objects can be measured. As in Fig 13, it is important to consider the likely error associated and it's effect on the bearing as the boat is moved. If you have latitude/longitude positions for your reference objects, by entering them as way-points, the GPS could be used to compute their bearings and distance without reference to a chart.

**The Sun or other astronomic object as a reference bearing.**
The sun, particularly when rising or setting, is often used as a reference bearing for compass checking. The traditional method is to compute its *amplitude*.

To use the technique you need to know your latitude and the sun's declination, which is obtainable from a Nautical Almanac. The traditional formula, which also works for the moon and other astronomic objects, is:

$$\text{Amplitude} = \text{Sin}^{-1}(\text{Sin(declination)} \div \text{Cos(Latitude)})$$

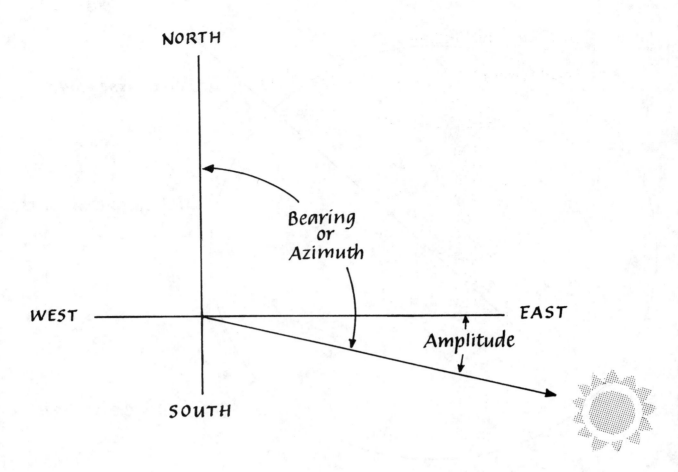

*Fig 14 Amplitude angles are measured from true east or west. Add or subtract from 90° or 270° to convert to a true bearing.*

Amplitudes are most effective within the tropics, particularly where your latitude is more or less equal to the sun's declination. Here, it changes little as it rises or sets, as its course across the sky is perpendicular to the horizon. In higher latitudes where it crosses the horizon at a small angle its amplitude changes quickly and the method is less reliable.

If you are familiar with astro/celestial navigation, these techniques can be used to find the sun or any astronomic object's azimuth, at any time. Use the Nautical Almanac to obtain the sun's Declination (Dec) and Greenwich Hour angle (GHA) for the required time. Enter this, along with your latitude and longitude, in the Sight Reduction Tables to obtain the sun's bearing or azimuth as it is known in this context.

Ideally, the calculation should be carried out for each azimuth used during the swing but because the calculation can be lengthy it is more convenient to draw up a table or graph of azimuth change throughout the whole time you anticipate the swing will take. In this way, the azimuth for any particular sight can be read off for the time in question.

By far the easiest way of obtaining the sun's azimuth is to use a computer program. Appendix 2 on page 53 gives notes for using and listings for a rudimentary program using an algorithm and sight reduction formula that are frequently used in astro/celestial navigation.

SUN POSITION PREDICTIONS          Position: 37 34 S 177 52 E

Local Date: 18 February 1998    Start Time: 0500 UTC

| Minutes | Alt.° | Azm.° | +180° | °/min |
|---------|-------|-------|-------|-------|
| +00 | 023 | 273 | 093 | -0.158 |
| +10 | 021 | 272 | 092 | -0.156 |
| +20 | 019 | 270 | 090 | -0.153 |
| +30 | 017 | 269 | 089 | -0.151 |
| +40 | 015 | 267 | 087 | -0.150 |
| +50 | 014 | 266 | 086 | -0.149 |
| +60 | 012 | 264 | 084 | -0.148 |
| +70 | 010 | 263 | 083 | -0.148 |
| +80 | 008 | 261 | 081 | -0.148 |
| +90 | 006 | 260 | 080 | -0.148 |
| +100 | 004 | 258 | 078 | -0.149 |
| +110 | 002 | 257 | 077 | -0.150 |
| +120 | -00 | 255 | 075 | -0.152 |
| +130 | -02 | 254 | 074 | -0.153 |
| +140 | -04 | 252 | 072 | -0.155 |
| +150 | -06 | 250 | 070 | -0.158 |
| +160 | -08 | 249 | 069 | -0.160 |
| +170 | -10 | 247 | 067 | -0.163 |

a)

Negative altitudes indicate that the sun is below the horizon

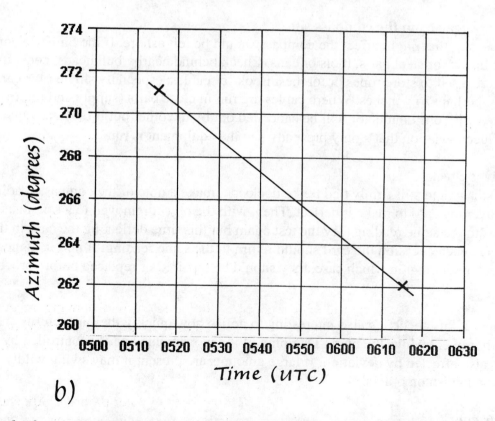

b)

*Fig 15 **Example of a sun azimuth table (from C-Swing) and graph for a swing carried out shortly before sunset in New Zealand***

39

### Use of a shadow pin

A shadow pin, no longer a common feature on marine compasses, is a vertical pin mounted in the center of the card, in line and above the pivot. In sunlight, the pin provides a shadow that falls across the card calibrations, forming a pointer to the sun's reciprocal bearing.

### How often should the compass be swung?

On any new boat or with a new compass installation, a swing to test for deviation is essential. There-after it should repeated as follows:

- At any time that you have reason to doubt the compass accuracy.
- If there has been a large (greater than 10°) change of latitude.
- Following a lightening strike
- Following any welding, installation of new equipment, major repairs or structural alterations
- After a long (greater than 6 months) lay-up period.

Fortunately, marine mines and armed hostilities are not a regular event for 'pleasure' craft since, as some military manuals advise, we would then need to add Degaussing and heavy gunfire to the list!

## Pre-swing check list

For compass swinging the boat should be prepared in its usual seagoing condition with all gear in place and electrical equipment functional. There are also some preliminary checks that help to ensure that the work to be carried out will not be wasted:

### 1) Remove stray gear from the compass site.

Any extraneous gear that might affect the compass should be left ashore. If the binnacle contains lockers, check inside for steel parts, tools or cans. Check behind nearby bulkheads, cockpit combing or canvas pockets used to store ropes or torches. Look for cables, especially single runs carrying heavy currents, perhaps to winches. Where cables are run in close pairs (supply and return) magnetic fields associated with one conductor will be cancelled out by the other but if they are separated, the fields can produce deviation that's only present when the equipment is run.

### 2) Sticking pivot check.

With the boat steady and still firmly tied to the dockside, make a note of the compass reading. Pass a steel object over the card to make it deflect. Then, with the object removed to a safe distance, see if the card returns to the same reading. Try the test again but this time deflecting the card in the opposite direction. On each occasion the card should return to the same reading. If not, it suggests that the pivots are worn or sticking, in which case, they should be repaired or replaced before proceeding further.

While carrying out this check it's also interesting to notice how quickly the card returns and the extent to which it over-swings the mark. Damping of the movement is partly controlled by design features but is also affected by deviation. If large amounts are present it may swing wildly on some courses and be sluggish on others.

### 3) Lubber line check.

Make sure the lubber line (vertical index line) on the forward side of the compass is parallel with the fore and aft line of the boat. If this is not done there will be a constant error on all compass readings.

When the compass is mounted on the centerline of the boat, check that it is aligned with the bow or some other point on the centerline. If it's offset, then measure the offset and make a mark at a point further forward with which to align it. (Points A & B in the drawing below)

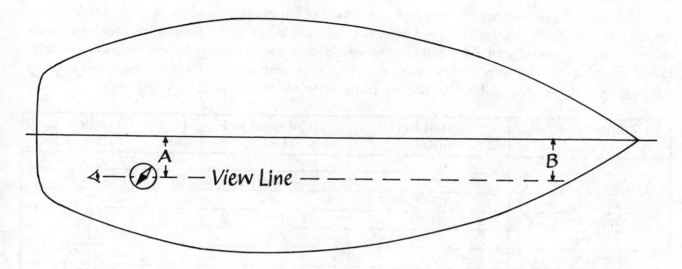

*Fig 16 Check that the lubber line is parallel to the boat's centerline.*

**4) A final check for iron or steel near the compass site.**
Don't forget to look through your own pockets for magnetic parts. Steel framed sunglasses, or perhaps steel buttons, a belt buckle or suspenders. It's amazing how easily such things can be overlooked but, if missed, the effect on the compass can be totally confusing. It can be infuriating to spend time correcting a compass only to find that a piece of iron in a binnacle locker affected the results. Have a last check for metal objects, cables behind paneling etc - not forgetting the bunch of keys in your pocket.

| | |
|---|---|
| Aerosol cans | Motors |
| Binoculars | Microphones |
| Cans of food or drink | Padlocks |
| Cassette players | Plastic parts with hidden steel reinforcing |
| Cell phones | Power cables (especially high current runs) |
| Echo sounders | Pump handles |
| Fire extinguishers | Radios |
| Hand bearing compasses | Radar sets |
| Hidden bolts or screws | Shackles |
| Hose clips | Steering chains |
| Instrument repeaters | Sun glasses |
| Keys in your pocket | Steel buttons or buckles |
| Knives | Tools in lockers |
| Loudspeakers | Winch handles |
| Magnetic types of stainless steel | Windscreen wiper motors |

Examples of items likely to cause deviation if within a metre of the compass site.

# Getting Started

Remember that, in essence, swinging the compass involves sailing the boat on a number of compass headings and using the same compass to measure the bearing of a reference whose correct magnetic bearing has already been determined. During the swing it is useful to have at least two people aboard. One person to manage the boat and one to take bearings and record the results. The headings usually chosen are the four cardinal (ie north, east, south and west) and four intercardinal points (ie northeast, southeast, southwest and northwest). Reasons for this will become evident in the next two chapters. Beginning with the boat sailing on, say, a northerly course, use the compass to measure the bearing of the reference object and begin filling in the following table:

| Compass Course (Ccse) | Actual Ref (mRef) | Compass Ref (Cbng) | Deviation (D) |
|---|---|---|---|
| North | 057° | 058° | -1 (1W) |
| North East | 057° | 060° | -3 (3W) |
| East | 057° | 059° | -2 (2W) |
| South East | 057° | 058° | -1 (1W) |
| South | 057° | 055° | +2 (2E) |
| South West | 057° | 053° | +4 (4E) |
| West | 057° | 055° | +2 (2E) |
| North West | 057° | 057° | Nil |

Calculate deviation from the definition on page 15 as follows:

**D** (Deviation) = **MRef** (Magnetic bearing of the reference object as obtained from the chart) - **Cbng** (Compass bearing of the reference object)

## Solutions to some practical difficulties

To use the compass you are swinging to measure bearings involves viewing the object across the centre of the card and reading its bearing directly from the scale. To achieve an acceptable level of accuracy, the compass must be fitted with rotatable sights and a bearing ring but few have these accessories. On other types such as those that are mounted against bulkheads or submerged in deep cockpits, the boat's structure may obscure the direct line of sight between the compass and reference.

There are several techniques that can be used to overcome this problem. One is to select a swing sight that has pairs of shoreside objects that can be used as transits on the four cardinal and four intercardinal points. This does not necessarily mean that you have to find 8 separate transits as each could be used as a back transit on a reciprocal course. Neither does it matter if they are not precisely on the cardinal/intercardinal points as the deviation on these points can be found by extrapolation and smoothing when the curve is drawn. What is important is that when the boat is on the transit it is not only sailing directly along it, but also heading in the same direction. Negligible wind and tidal streams are called for and a high level of boat handling skill.

## Use of a pelorus

A technique that's used to overcome this problem is to set up a remote compass card on a high part of the boat's structure. The card is fitted with its own sights and secured on an unobscured horizontal platform with it's 0° to 180° line running parallel to the boat's centerline. A remote card used in this way is known as a 'dumb' compass or *pelorus*. It measures the reference object's bearing from the boat's centerline; the same line from which the compass measures its magnetic heading. This is a method often used by professional adjusters from a Compass Dolphin (ie. a special sea mark erected specifically for compass correction). Here the Dolphin is used in transit with selected shoreside objects located on cardinal and intercardinal magnetic bearings.

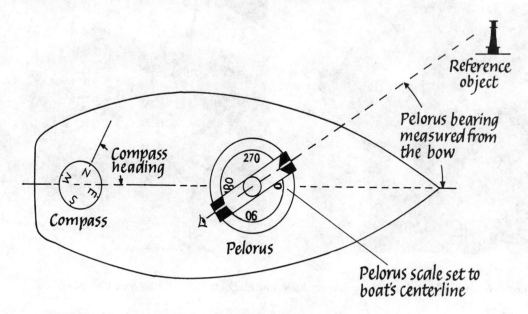

*Fig 17 **A pelorus set to measure angles from the boat's head.***

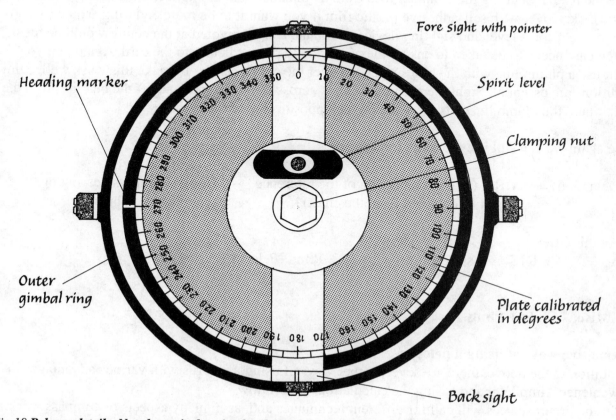

*Fig 18 **Pelorus details. Not shown is the tripod or base plate used to secure the instrument.***

*Photo 4 (a & b) a) Pelorus made by Davis Instruments b) Henry G. Dietz Pelorus and Sun Compass.*

If you don't actually have a pelorus, it is possible to make your own. In its simplest form you could make the plate by cutting the compass rose from an old chart and gluing it to a metal, plywood or plastic disk. Cover with self-adhesive plastic film if you want it to be more enduring. You will also need to add a spirit level and sighting device. A rotatable sight pivoted at the centre would be ideal though care needs to be taken to ensure they are exactly perpendicular to the card. At a pinch you might use a plumb line with the bob suspended over the card's center. To make this work well, calm conditions are essential. Sight the reference object across the scale and read off the bearing from the point where the plumb line cuts the scale on the opposite side of the card.

Once again we can still use the same formula :

**D** (Deviation) **=** **MRef** (Magnetic bearing of the reference object as obtained from the chart) **-** **Cbng** (Compass bearing of the reference object)

But in this case:
**Cbng =** **Chdg**(Compass heading) **+** **Pbng** (Pelorus bearing)

So:
**D = MRef - Chdg + Pbng**

### Alternative ways of using a pelorus
In addition to the above, there are several other ways of using a pelorus with various advantages in convenience, simplicity and easing the calculations that follow.
Different circumstances will call for different techniques and, as in many aspects of compass work, a certain flexibility of mind is needed to choose the best for the job in hand.

A popular method that helps to minimize the arithmetic is to use the pelorus as though it were a perfect, deviation free compass. Fig 19 sums up the situation. In this example the pelorus heading marker is set to 090° and sights are set to the magnetic bearing of the lighthouse. When the lighthouse appears in the sights the compass card should correspond to the pelorus exactly, showing that the boat is heading 090°M. Any difference is due to deviation.

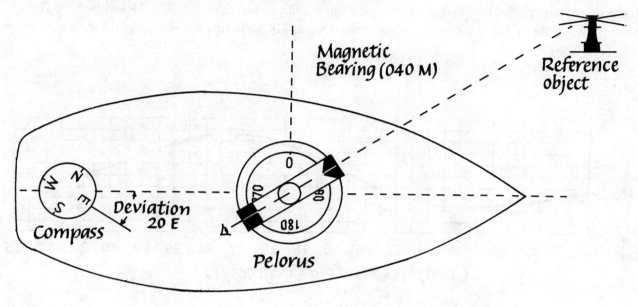

*Fig 19 Using the pelorus as a perfect deviation free compass.*

The entire swing can be carried out with the sights firmly clamped in the same position on the pelorus scale. With the boat sailing in a large slow circle, the scale and sights are rotated as one, so that the next cardinal or intercardinal point appears against the heading marker. Once again when the reference object falls exactly in the sights any difference between the compass and pelorus scales is recorded as deviation.

## Using a sextant

A marine sextant is the ideal instrument for accurately measuring angles under roughish conditions though it is not usually used in compass work. For this application, you will be using the sextant to measure the angle between your reference object and the boat's centerline. One way of doing this is to steer directly towards a distant object and measure the angle between it and your reference (not forgetting to correct the reading for the sextant's index error). Here, the main limitation on accuracy is likely to lie in your ability to steer an accurate course towards the chosen marker.

Alternatively, you may consider using a mark on the vessel itself, such as a light fitting on the bow. The line between the mark and axis of the sextant telescope should be parallel to the vessel's centre line. With the sextant mounted so that a line joining the telescope axis is exactly parallel to the centerline of the boat, the sight is taken by moving the index bar so that the reference object appears against the marker - just as you would bring the sun to the horizon in a conventional astro/celestial sight. Since the sextant telescope and index mirror are not on the same axis, to reduce parallax errors the sextant should be at least 10 metres away from the bow marker, which limits the use of this technique to larger boats only. A further inconvenience is that a conventional sextant is only capable of measuring angles up to 140° so the full horizon cannot be covered from a single station. This problem may be overcome by turning the sextant on its side to measure angles in the opposite direction or by using additional shore references or sighting stern marks from the bow.

**Processing the results**
Using graph paper, plot the deviation obtained during the swing as in Fig 20a. Draw the best-fit curve between them and use this to obtain deviations at 10° intervals for entering in the deviation table (Fig 20b)

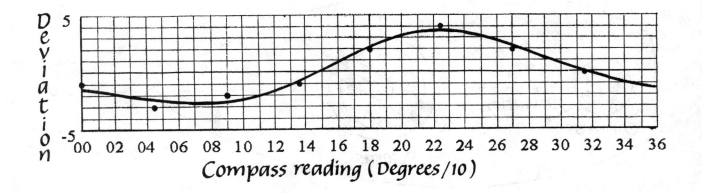

| DEVIATION TABLE | | | |
|---|---|---|---|
| **Heading** | **Deviation** | **Heading** | **Deviation** |
| North | 1W | South | 2E |
| 010 | 2W | 190 | 3E |
| 020 | 2W | 200 | 3E |
| 030 | 2W | 210 | 3E |
| 040 | 2W | 220 | 4E |
| 050 | 2W | 230 | 4E |
| 060 | 3W | 240 | 3E |
| 070 | 3W | 250 | 3E |
| 080 | 3W | 260 | 3E |
| East | 3W | West | 2E |
| 100 | 2W | 280 | 2E |
| 110 | 2W | 290 | 1E |
| 120 | 2W | 300 | 1E |
| 130 | 1W | 310 | Nil |
| 140 | Nil | 320 | Nil |
| 150 | Nil | 330 | 1W |
| 160 | 1E | 340 | 1W |
| 170 | 2E | 350 | 1W |

*Fig 20 After plotting the deviations, draw the best-fit curve between them and extract 10° values for the deviation table.*

**Computer assistance.**

If you have a computer, much of the arithmetic can be carried out for you. In the steps described so far, there is no great amount though a well designed program has the helpful advantage that the chances of trivial errors are very much reduced. Once set up, a simple spreadsheet for processing bearings would be useful for all subsequent compass swings. If your program is able to produce graphs, plotting the cardinal and intercardinal deviations may be possible and perhaps also extrapolation to produce a smoothed curve and table of deviations. Dedicated compass adjustment software such as C-Swing (see page 61), is also available and in addition to the ability to produce tables and graphs of deviations, may include curve fitting, sun almanac data and other mathematical tools.

**How much deviation is acceptable?**

If the deviation is never greater than '2° you need proceed no further. Simply keep the table in the boat's navigational area as a list of corrections to be applied on each heading.

Though it may not be possible to eliminate all deviation, any larger amount should be removed by *adjusting* the compass. This involves manipulating any internal adjusters provided with the compass or placing external magnets and correctors around the compass site in such a way that their magnetic effects are exactly equal and opposite to the boat's own field. On most small power and sailboats, this is a job that is well within the capabilities of the average owner and is the main subject for the next two Chapters.

# Chapter 5 - Analyzing Deviation

The magnetic field around a compass site may consist of a mixture of hard and soft iron effects that could be from a source below, above or to any side of the compass. Their combined influence may be extremely complex though fortunately it is possible to consider the result as due to the combined effects of no more than 5 separate components. These are called the *magnetic coefficients* and are identified as types **A**, **B**, **C**, **D** & **E**. In this chapter we look in detail at the cause and effects of each and the principles by which they are eliminated. Finally, the essence of the chapter is summarized in the table on page 41.

## Coefficient A    (effects all headings equally)

The effect of coefficient **A** is felt equally on all headings. It appears like a zero error where the correct compass reading can be obtained by simply adding or subtracting a constant amount, irrespective of the boat's heading. Fig 21 Shows a deviation graph for a compass affected by +3 degrees of type **A** deviation where:

### Deviation = A + Compass Heading

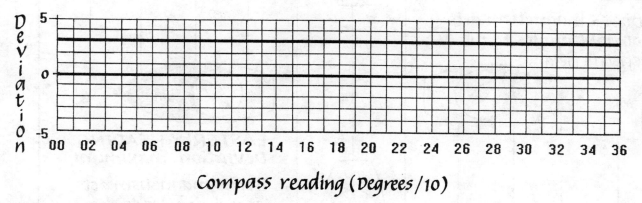

*Fig 21 Type A deviation (+3°)*

Though it is possible for type **A** deviation to have magnetic origins, this is unusual and a more likely cause is that the compass lubber line is not set parallel to the boat's centerline.

### Correcting A deviation

Simply slackening off the securing screws and rotating the compass body corrects the effect. In the above case, the compass is rotated +3° or, if you prefer, 3°E or 3° in a -clockwise direction. Bulkhead mounted compasses may present a special problem, particularly if the bulkhead is not at right angles to the boat's center line. In these cases it may be necessary to mount the compass against a specially prepared packing piece, with surface that's angled to correct the difference.

## Coefficient B    (greatest on Easterly and Westerly headings)

Let's suppose that a boat's bow behaves as a permanent South pole and the stern a permanent North pole. When the boat sails on a northerly or southerly compass heading the boat's magnetic field acts in exactly the same line as the earth's magnetic field. For the compass needle the field it experiences is either a little stronger or a little weaker, though its direction is not altered and no deviation appears.

Compare this with the situation when the boat is sailing on an easterly or westerly compass heading. In these cases, the boat's field is at right angles to the earth's field and becomes influenced by magnetism at the bow or stern. Unlike poles attract so the compass needle's south pole tends to be attracted to the stern and likewise it's north pole and the bow. It's exact angle is determined by the vector sum of the two components; a situation similar to that of a boat crossing a tidal stream, where it's actual course is determined by the vector sum of the boat's speed and heading and the speed and direction of the tidal stream.

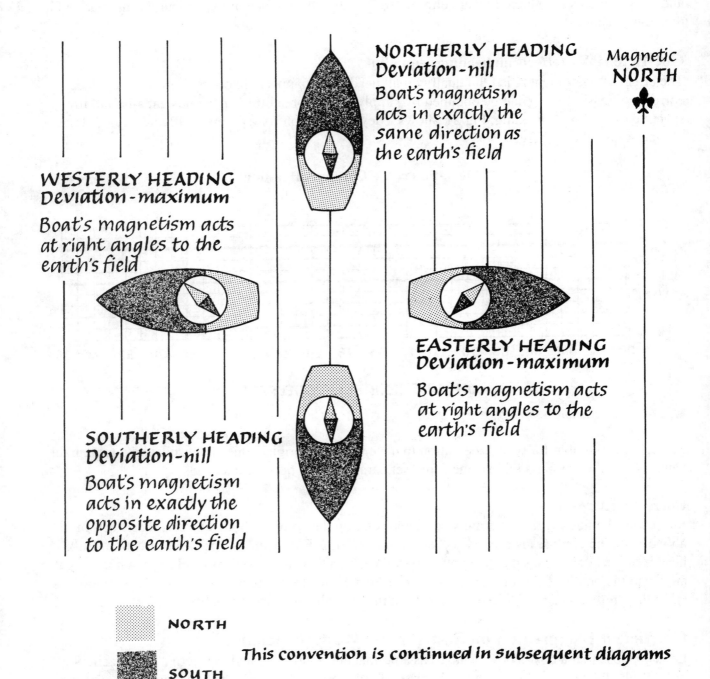

NORTHERLY HEADING
Deviation - nill
Boat's magnetism acts in exactly the same direction as the earth's field

Magnetic NORTH

WESTERLY HEADING
Deviation - maximum
Boat's magnetism acts at right angles to the earth's field

EASTERLY HEADING
Deviation - maximum
Boat's magnetism acts at right angles to the earth's field

SOUTHERLY HEADING
Deviation - nill
Boat's magnetism acts in exactly the opposite direction to the earth's field

NORTH

SOUTH

This convention is continued in subsequent diagrams

*Fig 22 Type B deviation caused by permanent magnetism at the bow and stern has maximum effect on east and west headings.*

This type of deviation is known as permanent **B** and is caused by hard iron magnetism running in the boat's fore and aft direction. This may be orientated with either the South (blue) pole forward (as in Fig 22) or the North (red) pole forward. In both cases the deviation graph follows a sine curve of the types shown in Fig 23 where

$$\text{Deviation} = B \times \text{sine(Compass Heading)}$$

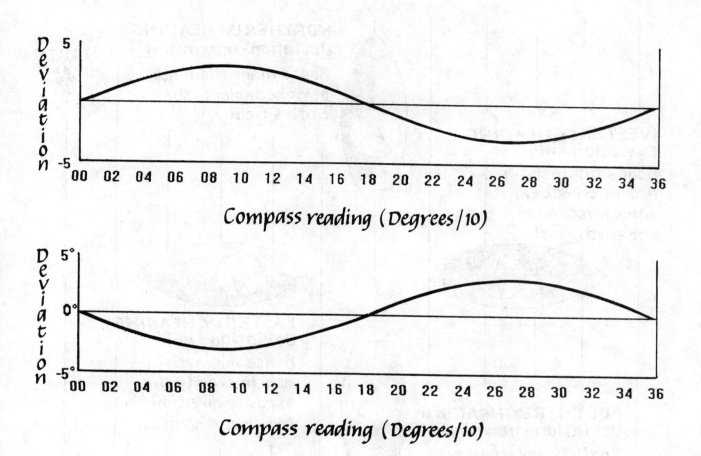

*Fig 23 Type B deviation curves. (a) Blue pole forward (+3°) (b) Red pole forward (-3°)*

## Correcting B deviation

Permanent **B** deviation can be eliminated by placing a small permanent magnet to the side of the compass site. It should run in the fore and aft direction and be orientated in the reverse direction to the boat's field. (eg for positive **B** deviation the corrector's red end should be forward).

Permanent **B** deviation is by far the most common on small boats and is often due to the motor, though a vertical mass of soft iron placed ahead or astern of the compass can produce a similar characteristic. This is known as induced **B** and was commonplace in the days when ships often had large steel funnels behind the wheelhouse. It was affected by latitude and corrected with a block of soft iron called a Flinder's Bar. Even for experienced adjusters, choosing the correct size and position often involved a good deal of guesswork.

## Coefficient C   (greatest on Northerly and Southerly headings)

Causes and characteristics of type **C** deviation are similar to type **B** except that the permanent magnetism lies athwartships rather than fore and aft. As a result, maximum deviation occurs on northerly and southerly headings. Taking an example where the blue pole is to starboard, deviations on the cardinal points are as follows:

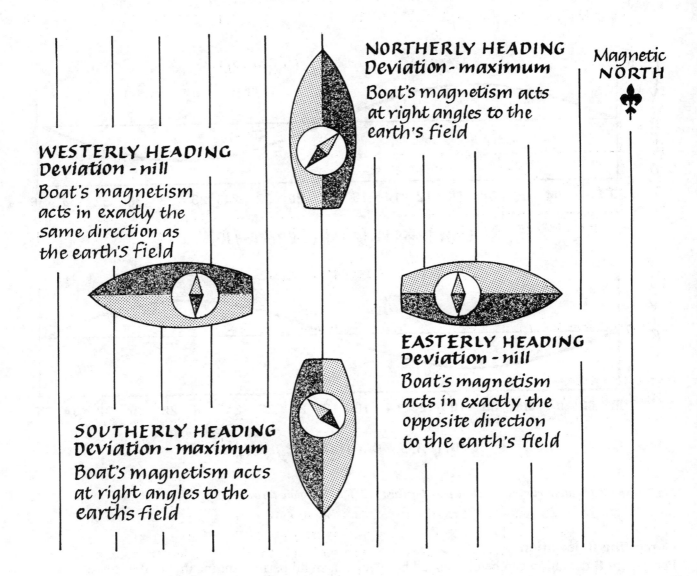

*Fig 24 Type C deviation caused by permanent magnetism to the side of the compass site has maximum effect on north and south headings.*

As with **B**, depending on the direction in which the boat's magnetism is polarized, the deviation graph will follow one of the two curves shown in Fig 25. Both are Cosine curves and:

$$\textbf{Deviation} = \textbf{C} \times \textbf{cosine(Compass Heading)}$$

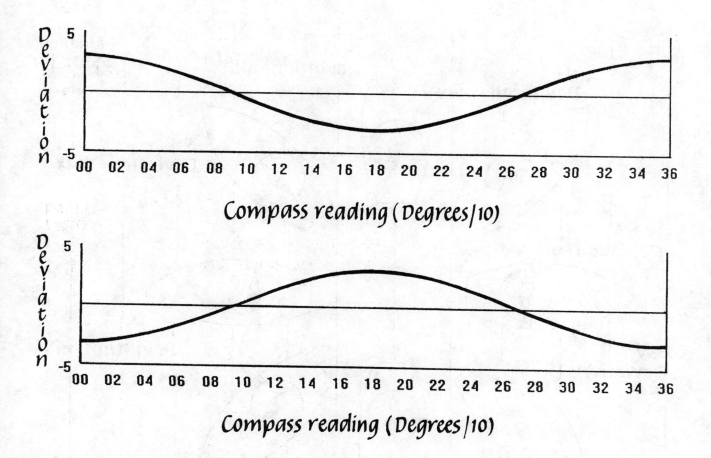

*Fig 25 Type C deviation curves. (a) Red pole to port (+3°) (b) Red pole to starboard (-3°)*

## Correcting C deviation

**C** deviation can be eliminated by placing a small permanent magnet ahead or aft of the compass site. It should be run in an athwartships direction and be orientated in the reverse direction to the boat's field. (eg for a positive **C**, the corrector's red end should be to starboard).

## Coefficient D   (greatest on NE, SE, SW & NW headings)

Unlike **B** & **C**, coefficient D is due to soft iron or temporary magnetism. This acquires its magnetic field from it's surroundings. Fig 26 illustrates the case of a boat with a symmetrical mass of soft iron running athwartships across the compass site which in
practice could be a cast iron engine block.  No matter which direction the boat is sailing, the iron forms a north pole at it's north facing end or side and it's effects on the compass needle can be predicted by applying the unlike poles attract rule. If the boat is sailing on north, south, east or west heading, no deviation occurs. On the four intermediate headings maximum deviation occurs, hence the name quadrantal error, which is given to errors of this type.

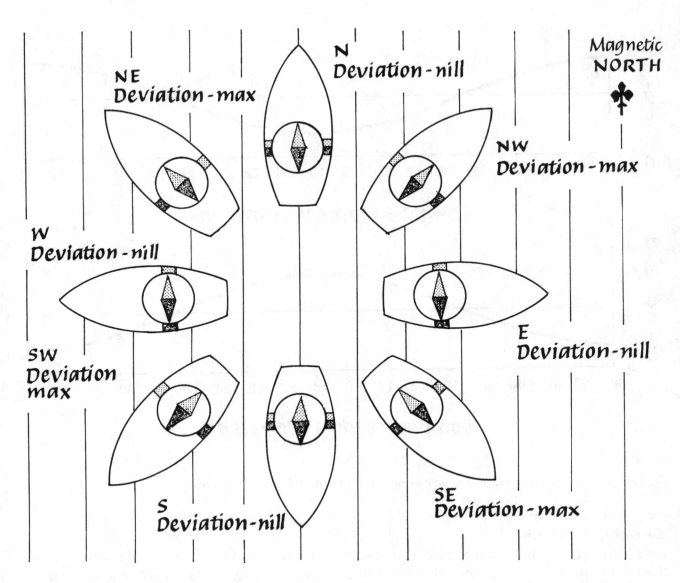

*Fig 26 **Type D deviation caused by soft iron temporary magnetism running across the compass site has maximum effect at the intercardinal points.***

In Fig 26 the iron mass is shown lying athwartships but note also that the same effect can occur for a similar iron mass lying in the fore and aft line. In the first case the deviation curve is positive and in the second negative. Both cases are shown in Fig 27 and are given by the relationship:

$$\text{Deviation} = D \times \text{sine}(2 \times \text{Compass Heading})$$

**Correcting D deviation**

**D** deviation is eliminated by placing blocks of soft iron, known as quadrantal correctors, at either side of the compass site. their shape is not important though usually they are spheres whose horizontal distance from the compass is adjustable. The deviation field produced by a single mass of iron running across the compass site has the same polarity as the field induced in the correctors. However, at the compass the correctors are able to cancel out the deviation since each produces a pole of opposite polarity next to the compass. Exact cancellation is achieved by carefully adjusting the distance between the correctors and the compass.

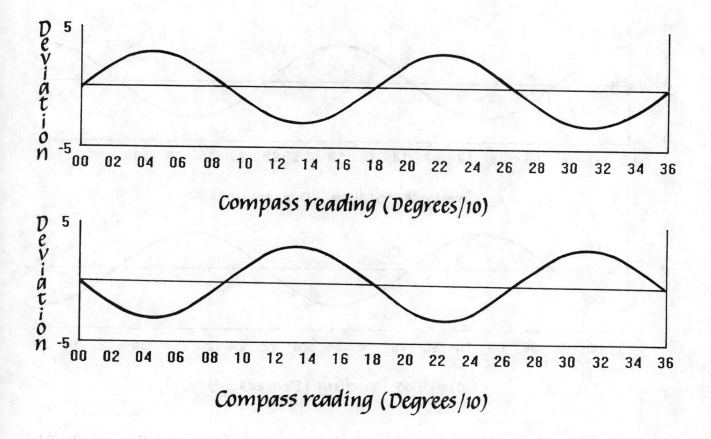

Fig 27 *Type D deviation curves. (a) Soft iron running athwartships (+3°) (b) Soft iron running fore & aft (-3°)*

## Coefficient E   (greatest on N, E, S & W headings)

Having got this far you can probably guess how coefficient **E** is derived. It's a soft iron effect exactly similar to **D**, except that in this case, the iron mass runs diagonally across the compass site instead of athwartships or fore and aft. Again, there are two possibilities; the iron may run starboard bow to the port quarter causing a positive **E** deviation, or  from the port bow to the starboard quarter causing a negative **E**. Both cases are shown in Fig 28 and are given by the relationship:

$$\textbf{Deviation= E x cosine(2 x Compass Heading)}$$

### Correcting E deviation

The same quadrantal correctors that were used to correct type **D** deviation can also be used to correct **E** though, in this case, they are rotated about the compass.

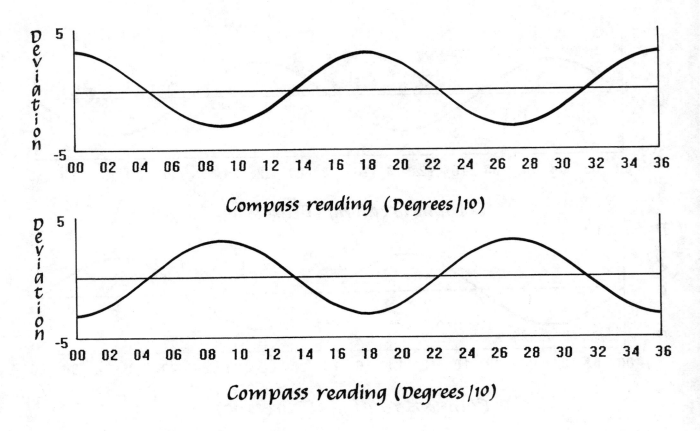

*Fig 28 Type E deviation curves. (a) Soft iron diagonally running from forward starboard to aft port (+3°) (b) Soft iron diagonally running from port forward to aft starboard (-3°)*

## Heeling error

So far we have considered deviation with the boat on a level keel, but what happens when it is heeled over? As it leans over the compass moves in it's gimbals and remains horizontal but some of the iron that was previously beneath it is now to one side, while other parts that were level with the card are now above or below it. The result is heeling error which is the difference between the vessel's deviation while upright and when heeled on the same heading.

Heeling error can be caused by soft or hard iron components and increases with the angle of heel (at least up to 45°). Likely causes are an engine mass or iron keel immediately below the compass site. An exact analysis of its causes and effects can be quite complex but in its most common form it has a maximum value when the boat is heeling to the east or west. This usually occurs when sailing north or south with the boat rolling or hard on the wind though it can occur with heavy pitching on an east west heading. With pitching or rolling its most noticeable effect is that the card oscillates excessively. Heeling error is dependent upon the angle of dip of the earth's field and without correctors, it increases with latitude.

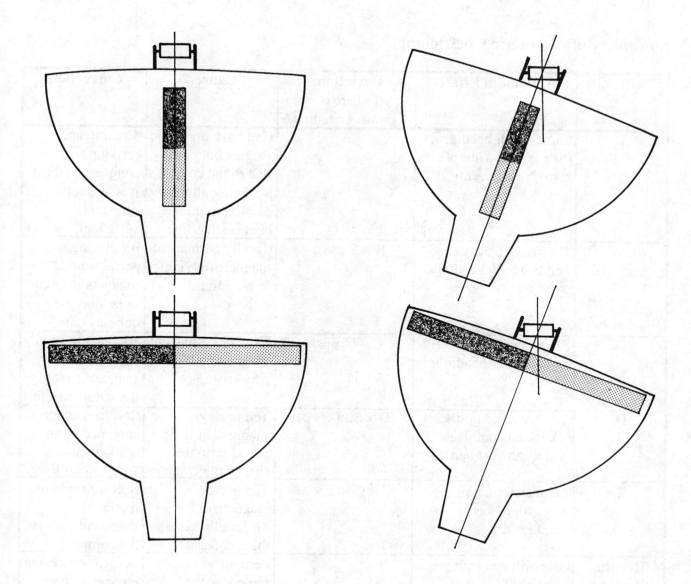

*Fig 29 **Changes in horizontal and vertical relationships between the compass and deviating fields as the boat heels.***

## Correcting heeling error

Heeling error is corrected with a vertical permanent magnet placed below the compass and in line with its center when the boat is not heeled. In the northern hemisphere the red end is placed uppermost. Adjustments are made by altering the magnet's size, and/or it's vertical distance from the compass.

# Summary of Magnetic Coefficients

| Coefficient | Maximum Effect | Deviation (where ø = compass heading) | Cause | Correction |
|---|---|---|---|---|
| A | Same on all headings. Like a like a simple zero or index error | A | Can have magnetic origins but usually due to the compass not being aligned parallel to the boat's centerline. | Rotating and refixing the compass so that it is correctly aligned is sufficient. |
| B | On magnetic easterly and westerly headings | $B \times Sin(ø)$ | Usually permanent magnetism lying fore-and-aft. | By placing small permanent magnets in front or behind the compass site. |
| C | On magnetic northerly or southerly headings | $C \times Cos(ø)$ | Usually Permanent magnetism lying athwartships. | By placing small permanent magnets to side of the compass site. |
| D | On magnetic northeast, southeast southwest and northwest headings | $D \times Sin(2 \times ø)$ | Temporary magnetism lying ahead or to the side of the compass | Soft iron spheres mounted in an athwartships or fore and aft line. |
| E | On magnetic north, east, south & west headings | $E \times Cos(2 \times ø)$ | Temporary magnetism lying diagonally across the compass. | Soft iron spheres mounted diagonally across the compass. |
| Heeling Error | On north and south headings | | Permanent or temporary magnetism. | Vertical permanent magnet placed below the compass |

## Calculating the coefficients

The deviation on any compass course (ø) is given by the expression:

**Deviation =
A + B × sine(ø) + C × cosine(ø) + D × sine(2 × ø) + E × cosine(2 × ø)**

This equation has 5 unknowns (the 5 coefficients) and can be solved by forming 5 simultaneous equations of observed deviations on 5 separate headings. There are several techniques that could be used though many would be extremely tedious to solve without a computer or least a calculator. Alternatively, the time honored method used by generations of compass adjusters requires only a paper, pencil, and the application of a few rules. It depends on using selected compass headings that reduce the sine and cosine terms to 0 or 1 which are the cardinal and intercardinal compass headings. The method works as follows:

1. Draw up a table of deviations as below.
2. Transfer all deviations to the **A** column naming westerly deviations as negative quantities. Find **A** by adding up the 8 deviations in this column and dividing by 8.
3. **B** deviation is a maximum on easterly and westerly headings where it is equal and opposite in sign. Transfer deviation on these headings from the **A** to the **B** column but change the sign of the second entry. Find **B** by adding up the 2 deviations in this column and dividing by 2.
4. **C** deviation is a maximum on northerly and southerly headings where it is equal and opposite in sign. Transfer deviation on these headings from the **A** to the **C** column but change the sign of the second entry. Find **C** by adding up the 2 deviations in this column and dividing by 2.
5. **D** deviation is a maximum on the intercardinal headings and on reciprocal headings is equal and opposite in sign. Transfer deviations on these headings from the **A** to the **D** column but change the sign on the second and fourth entry. Find **D** by adding up the 4 deviations in this column and dividing by 4
6. **E** deviation is a maximum on the cardinal headings and on reciprocal headings is equal and opposite in sign. Transfer deviations on these headings from the **A** to the **E** column but change the sign on the second and fourth entries. Find **E** by adding up the 4 deviations in this column and dividing by 4

| Compass Heading | Deviation | A | B | C | D | E |
|---|---|---|---|---|---|---|
| 000 | 10E | 10 | ~ | 10 | ~ | 10 |
| 45 | 20E | 20 | ~ | ~ | 20 | ~ |
| 90 | 11E | 11 | 11 | ~ | ~ | -11 |
| 135 | 1W | -1 | ~ | ~ | 1 | ~ |
| 180 | 5W | -5 | ~ | 5 | ~ | -5 |
| 225 | 8W | -8 | ~ | ~ | -8 | ~ |
| 270 | 12W | -12 | 12 | ~ | ~ | 12 |
| 315 | 7W | -7 | ~ | ~ | 7 | ~ |
| **Totals:** | | 8 | ~ | 15 | 20 | 6 |
| (average) **Coefficient:** | | +1 | 11.5 | 7.5 | 5 | 1.5 |

The value of coefficients obtained in this way is that they give an indication of the source of magnetism causing the deviation and provide a useful guide in placing the corrector magnets and soft iron adjusters needed to remove it. In the next chapter, we put theory into practice, look at compass corrector mechanisms, correction procedures and see how the coefficients are used as a methodical basis for eliminating deviation.

*Photo 5* **For training purposes, compass adjusters practice on a Deviascope. This is a flat wooden board fitted with a test compass and correctors. Additional hard and soft iron bars can be set to simulate various types of deviation and the whole instrument can be rotated and tilted to produce heeling effects.**

# Chapter 6 - Compass Adjustments

Having carried out a compass swing and found some significant amounts of deviation, the deviation curve may appear similar to the example in Fig 30a. In Fig 30b the curve is analyzed and **B**, **C**, **D** & **E** components are drawn separately. Were any **A** present, this would appear as a horizontal line. This is a typical small boat example where the contribution due to each is progressively less.

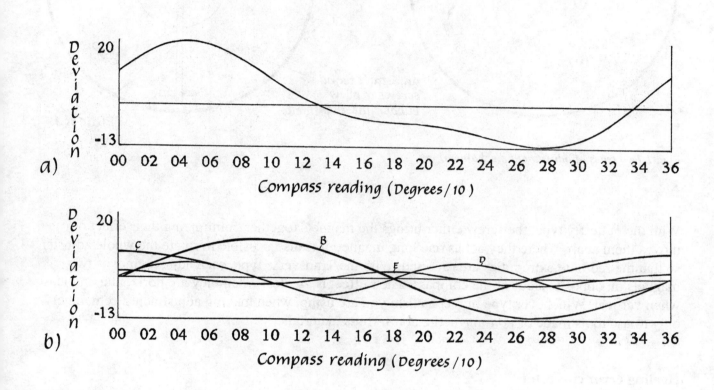

*Fig 30 (a & b)* **A typical deviation curve resolved into 4 separate constituents B, C, D & E.**

Before launching into the techniques used to reduce these, let's look first at the permanent magnets and soft iron temporary magnets that are used as controls for adjustments. Many, particularly better quality compasses, are fitted with internal permanent magnet adjusters for eliminating **B** and **C** deviation. Details vary, though the principle remains the same. In some, they are simply horizontal pockets or tubes fitted to the base of the instrument into which one or more magnets can be placed. In others, there may be screw adjusters that alter the relative position of pairs of correctors. Fig 31 **Two types of adjustment mechanism.** show examples.

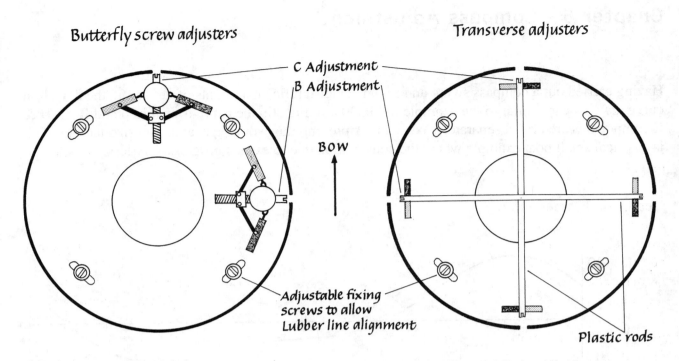

Fig 31 *Two types of adjustment mechanism.*

With the butterfly type, the screw either brings the magnets together, minimizing their effect or moves them apart, where they act as one long magnet. It is also possible to rotate the whole adjuster so that the red/blue extremities are reversed. With the Transverse type, each adjuster moves two magnets on opposite sides of the compass. Their effect is greatest when they are horizontal and least when vertical. Whichever type of mechanism you are using, when making adjustments, remember only to use tools made of non-magnetic, non-ferrous materials.

### Heeling error corrector
This permanent magnet corrector is usually hidden within the compass enclosure or binnacle and lies immediately below the compass but not sharing the same gimbal.
It may consist of a 'bucket' used to hold a number of magnets, in which case they should be distributed symmetrically about the compass centerline. Alternatively, the magnet may be raised or lowered on a threaded bar.

### External adjusters
Where no internal adjusters are fitted or if they are of insufficient strength, external permanent magnet adjusters can be used. For convenience, these are usually fitted below the card level on or below the surface on which the compass is mounted. These provide a very flexible and wider range of adjustment than can be achieved with internal magnets. For this reason adjusting with external magnets is often easier and more intuitive than with internal correctors. The downside is that once the job is done, the external magnets may not look as neat as a hidden internal adjuster and their need to be secured and protected from weather and corrosion.

The effect of external magnets can be altered by:

1. Reversing their direction.
2. Moving the magnet either closer or further away from the compass.
3. Using a longer or shorter magnet.
4. Placing additional magnets alongside.

In general, a longer, stronger magnet that's further away has a better characteristic than a smaller magnet placed closer to the compass.

## Quadrantal correctors

The single or paired iron spheres used to correct **D** and **E** deviation (quadrantal errors) are always external adjustments. Their diameter is normally between 40mm and 200mm and one or a pair may be fitted to brackets either side of the compass. Ideally, any soft iron corrector should be free from permanent magnetism. If any is present, it becomes a source of **B** and **C** deviation. As a test, move the spheres close to the compass body and slowly rotate each in turn. Any residual magnetism causing a deflection of more than 2° should be removed by annealing. The process involves heating the corrector to a temperature of around 700°C (dull red) and allowing it to cool slowly over a period of 10 hours. An easy way to do this, if you happen to have one available, is with a blacksmith's forge where the iron is submerged in the fire which is allowed to die overnight.

## Interactions between correctors

In the explanation of deviation coefficients given so far, we have assumed that each can be corrected independently of the others. Unfortunately, as there is always some interaction between correctors, this is not usually possible in practice. Soft iron correctors, for example, can acquire magnetism from permanent magnets used to correct **B**, **C**, heeling error or even the compass needle.

To cope with these difficulties, a methodical approach is essential, beginning with those corrections that are most likely to have the largest interactive effects. The recommended sequence is as follows:

| Order | Corrector | Carried out at |
|---|---|---|
| 1 | Lubber line alignment | Dockside |
| 2 | Flinder's bar | Dockside |
| 3 | Quadrantal correctors | Dockside or at sea |
| 4 | Heeling error magnets | Dockside or at sea |
| 5 | Fore & Aft magnets | At sea |
| 6 | Athwartships magnets | At sea |

Table 6.1 Recommended sequence of adjustments.

## Frozen Needle

In extreme cases where the boat has particularly strong permanent magnetism, its influence may overcome the directive force of the earth's field causing the compass needle to become sluggish in some sectors or perhaps totally *'frozen'*. It sticks close to a single reading regardless of heading and it turns with the boat. The solution is to place a permanent magnet to oppose and neutralize the strong boat field; the frozen needle indicates the alignment and polarity of the corrector. This is most easily done by setting the boat on the heading shown by the frozen needle, when the effect of the corrector (usually a **B** or **C** type) will be immediately obvious. Only when this is carried out and the needle is able to move freely, can other adjustments be made.


## Carrying out the adjustments

Before beginning any compass work, first work through the 4 point *Pre-Swing Check list* given on page 30. If the compass or boat is new, remove all adjusters or set their effect to a minimum. If any **A** deviation is present, this could be due to some miscalculation in handling bearings during the swing. Magnetic causes are a possibility, though it is more likely that the lubber line is misaligned. In either case, the error can be eliminated by rotating the compass body (see page 31). This and several other parts of the adjustment are best carried out with the boat still tied to its berth. Clearly, it is a good idea to complete as much as possible on the dockside as you are spared the pressures of having to manage the boat and can work at your own speed. With a suitable swing site, the adjustments listed in table 6.1 can now be completed. In principle, one could carry out a full compass swing between each stage of corrections. This would give a good picture of the effect of each correction and is a good exercise but because of the time taken this is seldom done in practice. For most small boats the work can be cut down to size by making the corrections as follows and then taking a test swing as an overall check on the results.


### (1) The Flinder's Bar

This is used to correct induced **B** and **C** deviation which, as mentioned in Chapter 5, was once used to correct for the effects of a ship's funnel. A steel mast could produce a similar effect, though on today's small craft, this type of structure is unusual. Calculating or estimating the length of bar is a job that requires experience with vessels of a similar type and is a good reason for using the services of a professional adjuster. For this reason and because Flinder's bars are seldom necessary, we will not consider them further but move on to the next adjustment.


### (2) Quadrantal Correctors

Because **D** and **E** deviation is less likely on boats made of nonferrous materials, these are generally only fitted to compasses on steel or ferro-cement hulls. If these errors are present, set the boat on an intercardinal heading and slacken the bolts holding the correctors. Move them in an athwartships direction, towards or away from the compass site until the **D** error is just removed and retighten the bolts. Remember that this is only part of the total error and is likely to be less than that due to **B** or **C**.

**E** deviation is comparatively rare but if present, place the boat on a cardinal heading and remove it by slewing the correctors around the compass site. Correcting for **D** or **E** on a single intercardinal or cardinal heading also corrects the error on all others.


### (3) Heeling Error

Because they spend a good deal of time heeled over this is more important on monohull sailing boats rather than power boats. None the less, it is a correction that can be difficult to achieve and some professional correctors prefer simply to advise navigators to be aware of it's existence and that it can change with latitude.

If a large amount of heeling error is believed to be present (see page 40) you may consider enlisting the help of a professional adjuster. Clearly, particularly with large vessels, there could be great practical difficulties if it were necessary to experimentally heel the boat by measured amounts. Instead, the conventional method of correction is by using a Vertical Force Instrument (see Photo 6) that can be carried out with the boat on an even keel at the dockside.

*Photo 6 A Vertical Force Instrument used to correct heeling error.*

One commonly used type includes a horizontally pivoted magnetic needle that responds to the vertical or dip component of the earth's magnetic field. Set against it is a scale indicating deflection from the horizontal and attached to the needle itself is an adjustable weight. This allows the needle to be balanced in a horizontal position regardless of the local angle of dip. Before use, it must be set in this position with the instrument ashore and in a region free from interfering magnetic fields. While this is being done, it should be in a level position and turned so that the north end of the needle is towards the north.

In use, the compass is removed and the instrument set in its place so that its needle occupies the same position as the compass needle. Again, the north end should be pointing towards the north. In this position, any deflection of the needle from the horizontal is removed by moving the heeling error magnet up or down or changing the number of magnets. Ideally, for this adjustment, the boat should be set on an east or west magnetic heading. The method as described is suitable for boats where the compass is not enclosed in a steel wheelhouse where additional corrections are required.

### An Alternative Method

Not many boat owners have access to a vertical force instrument. However, there is a less convenient alternative that can be used to correct the error at sea. Choose a day with a fairly lumpy sea but little wind; perhaps following a few days of strong winds. With the boat on a north or south heading encourage it to roll heavily by not setting sails and observe how the compass card swings. Adjust the heeling error corrector so that the card's oscillations are at a minimum.

Before leaving the subject of heeling error, remember that in the northern hemisphere the red end of the heeling error corrector should be placed uppermost, and that after a significant (more than 10°) change of latitude it may need to be readjusted.

### (4) Fore & Aft Magnets

These are used to correct for **B** deviation which, together with **C**, are the most significant for small boats, regardless of the hull materials. If **B** error is present, sail on an easterly or westerly heading and manipulate the adjuster so that the error is just removed.

If you are using external corrector magnets, begin by holding the corrector in a fore and aft direction but well away from the compass. Bring it slowly towards the compass side watching how the error changes. If it increases then reverse the magnet's polarity. Find a position where the error is just removed then fix the magnet temporarily in place with double or single sided adhesive tape or 'Blu Tack'.

### (5) Athwartships Magnets

To remove **C** deviation, follow the same procedure as above by adjusting the athwartships corrector. If you are using external correctors, the procedure is the same except that the magnet is aligned athwartships and placed ahead or aft of the compass.

## The Short Swing Method

The above method (sometimes referred to as the Analysis Method) is fairly labor intensive requiring a complete compass swing, analysis and correction of the coefficients followed by a final swing to make sure that all has gone well. It has the advantage that it is fairly foolproof. As a practical alternative for most small boats, it is possible to make assumptions about the type of deviation and achieve several short cuts.

To begin, let's imagine that any **A** deviation can be eliminated by rotating the lubber line adjustment. Let's also assume that all **B** and **C** deviations are due to permanent magnetism. Because it's generally only found in small amounts and then only in rare cases, **E** can also usually be ignored. Now look again at Fig 30b and imagine the **E** component removed. This leaves **B**, **C** and **D** of which **B** is the most significant. Notice that on east and west headings this is now the only component present, since **C** and **D** are zero. Similarly, notice that on north and south headings **C** is the only component present.

These principles form the basis of the widely used short swing method. For convenience, this is often used with a selection of reference objects close to cardinal and intercardinal magnetic headings.

## 1) Correct for B deviation

Choose a transit that runs in an approximately east to west (magnetic) direction. With the boat heading west along the transit, make a note of the exact compass reading. Next, turn the boat and steer in the opposite direction along the same transit and again make a note of the exact compass reading.

After subtracting 180° from the reading on the westerly heading, find the average of the two readings which, since the deviation on each heading is of equal and opposite sign, should be the magnetic bearing of the easterly leg of the transit. Check this from the chart if you are unsure. Now calculate the deviation on each heading from the following expression:

**Deviation = Magnetic Heading - Compass Heading** (see page 15)

**For Example:**

| | | |
|---|---|---|
| Steering directly towards a westerly reference gives a compass reading of: | 275 | |
| Subtract 180°: | | 095° |
| On a reciprocal course with the same object astern the compass reads: | 073 | 073° |
| Sum: | | 168° |
| Divide by 2 to find the actual magnetic bearing of the object: | | 084° |
| Deviation on the westerly heading = 084 + 180 - 275 | -11 | 11W |
| Deviation on the easterly course = 084 - 073 | +11 | 11E |

Remove the deviation by again setting the boat on one of the two headings and manipulating the fore and aft **B** adjuster as described for the previous method.

## Note

With the short swing method, the possibility of other forms of deviation being present cannot be totally eliminated. When correcting for **B,** for example, it is conceivable that some magnetic **A** may be present. For this reason, it is common practice to only remove half of the indicated deviation.

This strategy helps prevent over correction and the possibility of entering an endless cycle of correction, swing and further correction.

## 2) Correct for C deviation

The procedure here is exactly the same as for correcting **B**, except, of course, that you will need a reference object or transit that gives a magnetic north or south heading. In finding the average of the two headings, decide when to add or subtract 180° in order to get a sensible answer.

**For Example:**

| | | |
|---|---|---|
| Steering directly towards a northerly reference gives a compass reading of: | 006 | |
| Subtract 180°: | | -174° |
| On a reciprocal heading with the same object astern the compass reads: | 160 | 160° |
| Sum: | | -014 |
| Divide by 2 to find the actual magnetic bearing of the object: | | -007° |
| | | |
| Deviation on the northerly heading = -007 - 006 | -13 | 13W |
| Deviation on the southerly  course = -007 +180 - 160 | +13 | 13E |

Similarly to the **B** adjustment, **C** deviation is removed by again setting the boat on one of the two headings and manipulating the athwartships adjuster as described previously.

## 3) Correct for D deviation

Because quadrantal deviation in opposite quadrants is of the same sign, the reciprocal heading technique used for **B** and **C** cannot be used for **D**. Instead, the boat is put on a known intercardinal magnetic heading. Deviation is calculated as usual by subtracting the compass reading from known heading. Any deviation is just eliminated by moving the quadrantal corrector(s) closer to or further away from the compass, which also corrects **D** on all other intercardinals.

## Finishing the Adjustment

At this stage, make a further check to ensure that all spare magnets have been cleared away and that during the preceding work no steel objects have been left near the compass site. It's amazing how easily commonplace magnetic objects can be overlooked, though their effects may render your efforts worthless.

To make sure that all has gone well, carry out a final swing to confirm the adjustments. Hopefully, the result will be a chart with little (less than 2°) or no deviation and the satisfaction of a job well done. Just possibly (through interaction of adjusters), you will need to recorrect on some headings though to aim for nil deviation all around the compass may be unrealistic. For future reference, in case they are disturbed later, record on the deviation chart the position, number and orientation of all correctors used.

## Securing the Adjusters

With some types of internal adjusters, it is possible for vibration to loosen the mechanism and over a period of time adjustment is lost. Consider using a thread locking compound or a dab of marine sealant if this seems likely. With magnets that are held in tubes, make sure that they are well sealed and that water cannot enter and start corrosion.

If you have been using external permanent magnet correctors, their positions during the correction were fixed only temporarily. To make a more permanent installation they need to be fixed securely and provided with protection from the weather and corrosion. Some types are supplied in sealed brass tubes with drilled lugs for screw attachment. Don't forget to use non-magnetic screws; even chromium-plated screws are often made of steel. If you are using plain bar magnets, a split plastic pipe as shown in Fig 32 can provide a convenient cover.

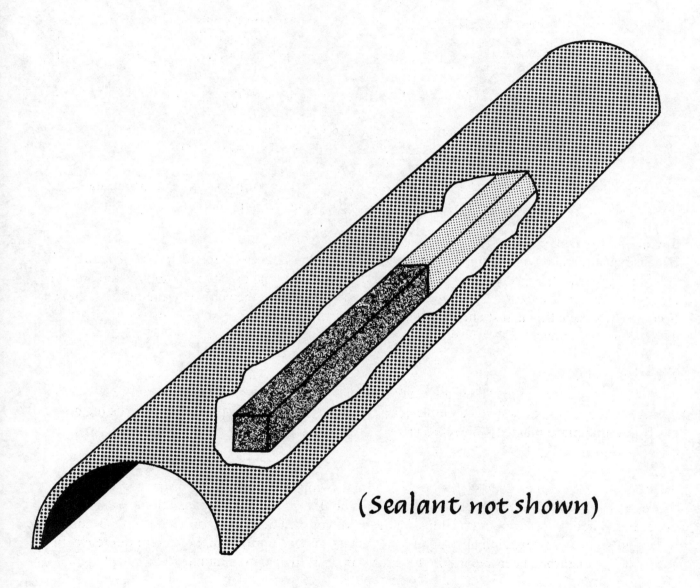

(Sealant not shown)

*Fig 32 A short length of split plastic water pipe can be used to both secure and protect corrector magnets. With the magnet fixed temporarily with adhesive tape, fill the half-round section with marine sealant and place it over the magnet so that the excess squeezes out. Wipe off the excess and leave to cure.*

# Appendix Notes

## Appendix 1 - Computer Software

### Geomag - Geomagnetic field modeling

If you are using a computer, a further source of this information would be to use a geomagnetic model program such as Geomag. At the time of writing, this could be obtained from the Geophysical Data Center's web site (see page 61). Geomag is not an official product so is not recommended for critical applications though it is valid for several years and able to produce a range of magnetic data for any geographic location.

### MagVar - Magnetic Variation Applet

A java applet that includes a Mercator projection world map. Moving the mouse cursor over the map gives a continuous readout of latitude/longitude and the corresponding variation. (www.pangolin.co.nz)

### C-Swing - Software Compass Adjuster

A Windows 95 or Windows NT program for handling much of the routine 'book keeping' work of small boat compass adjustment. Its onboard math package has the following capabilities:

*   Handles bearings taken directly or by pelorus.
*   Computes the sun's azimuth in real time for any geographic location.
*   Handles reference bearings from any direction (not just cardinals and intercardinals).
*   Computes the 'best fit' deviation curve from observed data.
*   Computes coefficients **A, B, C, D & E**.

Supplied through: Pangolin Communications (see Appendix 4 - Suppliers and Web sites)

# Appendix 2 - A Sun's Azimuth Program

Several long-term algorithms exist for computing the position of the sun. On the whole, the precision improves as the method becomes more complex but for compass work ±0.1° is usually sufficient. If the mathematics extends for no more than a few lines, solution by pocket calculator may be possible; if you have a computer language at your disposal, a short program is a more convenient option.

Two sets of listings for SunPos, a sun's position program, are given below. These are written in Basic and C++ using the same variable names, methods and program flow. Both use only simple statements that are common to most other computer languages so conversion to another one of your choice should be fairly straight-forward. The output provides the predicted position of the center of the sun's disk. The altitude given is uncorrected for refraction which varies from zero, when the sun is immediately overhead, to around half a degree when it is close to the horizon.

To test the program, try the following example:

Position:     50°10'N 1°20'W
UTC Date:   August/23/1999
UTC Time:   14:20

As either version of SunPos runs through, it begins by asking for a position, date and time as follows:

Latitude?     50.1667      (degrees and decimals - southern latitudes are negative)
Longitude?   -1.333        (degrees and decimals - western longitudes are negative)
Month?        8
Day?           23
Hour?          14
Minute?        28
Altitude =     41.83
Azimuth =     228.96

In this example, the year was set at 1999. Predictions for other years can be computed by substituting other values for the sun's Mean Anomaly and Earth's longitude at perihelion. This data is given for the first decade of the 21st century in table A2 and should be inserted into the program variables Mo and Lp for the year in question. For convenience you could use a CASE or DATA statement in Basic or switch statement in C++.

To avoid non essential clutter, SunPos is given in a fairly rudimentary form. For convenience you use CASE, DATA (Basic) or switch (C++) statement to load Mp and Lp automatically. By using a loop, the program could produce a table of azimuths at, say, 10 minute intervals from a specified start time.

| Year | Mo | Lp |
|------|--------|---------|
| 1997 | -3.1930 | 77.1087 |
| 1998 | -3.4505 | 77.0916 |
| 1999 | 3.7078 | 77.0744 |
| 2000 | -3.9523 | 77.0605 |
| 2001 | -3.2217 | 77.0411 |
| 2002 | -3.4828 | 77.0236 |
| 2003 | -3.7354 | 77.0092 |
| 2004 | -3.9908 | 76.9909 |
| 2005 | -3.2630 | 76.9706 |
| 2006 | -3.5156 | 76.9577 |
| 2007 | -3.7738 | 76.9424 |
| 2008 | -3.0292 | 76.9231 |
| 2009 | -3.3014 | 76.9041 |
| 2010 | -3.5568 | 76.8858 |

Table A.2

# SunPos Program Listings in C++

```cpp
#include <iostream.h>
#include <math.h>

//Use data from table A.1 to set year Mo and Lp
int Year=1999;
double Mo=-3.7078;
double Lp=77.0744;

#define PI 3.1415926535

//Define trig functions that work with angles in degrees
inline double sind(double x){ return(sin(PI*(x)/180));}
inline double cosd(double x){ return(cos(PI*(x)/180));}
inline double tand(double x){ return(tan(PI*(x)/180));}
inline double dasn(double x){ return(180/PI)*(asin(x));}
inline double dacs(double x){ return(180/PI)*(acos(x));}
inline double datn(double x){ return(180/PI)*(atan(x));}
inline double dunw(double x){ return(x - (floor(x/360)*360));}


//Function prototypes
double DayNumbers(int Year, int Month, int Day, int Hour, int Minute, int Second);
void   Sun(double DayNo, double *Gha, double *Dec);
void   SightReduction(double Lat, double Lng, double Dec, double Gha, double *Altitude, double *Azimuth);
```

74

```cpp
void main(){
//Main program starts here
        int Month;      int Day;
        int Hour;       int Minute;
        double Lat, Lng, DayNo, Gha, Dec;
        double Altitude, Azimuth;

//Get initial data
        cout << "Latitude? ";  cin >> Lat;
        cout << "Longitude? "; cin >> Lng;
        cout << "Month? ";      cin >> Month;
        cout << "Day? ";        cin >> Day;
        cout << "Hour? ";       cin >> Hour;
        cout << "Minute? ";    cin >> Minute;

//Call functions
        DayNo=DayNumbers(Year, Month, Day, Hour, Minute, 0);
        Sun(DayNo, &Gha, &Dec);
        SightReduction(Lat, Lng, Dec, Gha,
    &Altitude, &Azimuth);
//Print results and finish
        cout << "Altitude  = "<<Altitude << "\n";
        cout << "Azimuth = "  <<Azimuth  << "\n";
}
//————————————————————————————————//Compute number of days
and fractional days from the //beginning of the year
double DayNumbers(int Year, int Month, int Day, int Hour, int Minute, int Second){

        double yr = Year;
        double mn = Month;
        double dy = Day;
        double hr = Hour;
        double mi = Minute;
        double se = Second;
        double y = 0, d = 0;

        if(floor(yr/4) == yr/4) y = 1;
                if(mn > 2){
                d = floor(30.6 * (mn+1))-63+y;
        }else{
                d = floor((63-y)*(mn-1)/2);
        }
        return (d+dy+(hr/24)+(mi/1440)+(se/86400));
}
//————————————————————————————————//Compute sun's Declination
and Greenwich hour angle
void Sun(double DayNo, double *Gha, double *Dec){
```

```
        double u, v, x;

        u = (.9856*DayNo)+Mo;
        v = u+(1.916*sind(u))+(.02*sind(2*u))-Lp;
        *Dec = dasn(.3978*sind(v));
        x = datn(.9175*tand(v));
        if((sind(x) > 0) != (sind(v) > 0)) u = u+180;
        *Gha = 360*(DayNo-floor(DayNo))+u-x-180-Lp;
}
//————————————————————————————————
//Convert co-ordinates from ecliptic (Dec & GHA) to //observer centered(altitude & azimuth).
void SightReduction(double Lat, double Lng, double Dec, double Gha, double *Altitude, double
*Azimuth){

        double Lha, tmp;

        Lha = dunw(Lng+Gha);
        tmp=sind(Lat)*sind(Dec)+cosd(Lat)*cosd(Dec)*cosd(Lha);
        *Altitude=dasn(tmp);

        tmp = sind(Dec) - sind(*Altitude)*sind(Lat);
        tmp = tmp/cosd(*Altitude)/cosd(Lat);
        *Azimuth = dacs(tmp);
        if(Lha < 180) *Azimuth=360-*Azimuth;
}
//————————————————————————————————
```

# SunPos Program listings in Basic

```
'Use data from table A.1 to set year, Mo and Lp
Year = 1999
Mo = -3.7078
Lp = 77.0744

PI = 3.1415926535#
'Main program starts here
'Get initial data from the user
INPUT "Latitude? ", Lat
INPUT "Longitude? ", Lng
INPUT "Month? ", Month
INPUT "Day? ", Day
INPUT "Hour? ", Hour
INPUT "Minute? ", Minute

'Call sub-routines
GOSUB DayNumbers
```

```
GOSUB Sun
GOSUB SightReduction

'Print results
PRINT "Altitude  = "; Altitude
PRINT "Azimuth  = "; Azimuth
'All finished
END
'_____

'Compute number of days and fractional days from the
'beginning of the year
DayNumbers:

IF (INT(Year / 4) = Year / 4) THEN y = 1
IF (Month > 2) THEN
        b = INT(30.6 * (Month + 1)) - 63 + y
ELSE
        b = INT((63 - y) * (Month - 1) / 2)
END IF
DayNo = b + Day + (Hour / 24) + (Minute / 1440) + (Second / 86400)
RETURN
'_____

'Compute sun's Declination and Greenwich hour angle
Sun:

u = (.9856 * DayNo) + Mo
v = u + (1.916 * sind(u)) + (.02 * sind(2 * u)) - Lp
Dec = dasn(.3978 * sind(v))
X = datn(.9175 * tand(v))
IF ((sind(X) > 0) <> (sind(v) > 0)) THEN u = u + 180
Gha = 360 * (DayNo - INT(DayNo)) + u - X - 180 - Lp
RETURN
'_____

'Convert co-ordinates from ecliptic (Dec & GHA) to
'observer centered (altitude & azimuth).
SightReduction:

Lha = dunw(Lng + Gha)
Altitude = dasn(sind(Lat) * sind(Dec) + cosd(Lat)·* cosd(Dec) * cosd(Lha))
Azimuth = dacs((sind(Dec) - sind(Altitude) * sind(Lat)) / cosd(Altitude) / cosd(Lat))
IF (Lha < 180) THEN Azimuth = 360 - Azimuth
RETURN
'_____

//Define trig functions that work with angles in degrees

FUNCTION sind (X)
sind = SIN(4 * ATN(1) * (X) / 180)
```

```
END FUNCTION

FUNCTION cosd (X)
cosd = COS(4 * ATN(1) * (X) / 180)
END FUNCTION

FUNCTION tand (X)
tand = TAN(4 * ATN(1) * (X) / 180)
END FUNCTION

FUNCTION dasn (X)
dasn = (180 / (4 * ATN(1))) * (ATN(X / SQR(-X * X + 1)))
END FUNCTION

FUNCTION dacs (X)
dacs = (180 / (4 * ATN(1))) * (4 * ATN(1) / 2 - ATN(X / SQR(-X * X + 1)))
END FUNCTION

FUNCTION datn (X)
datn = (180 / (4 * ATN(1))) * (ATN(X))
END FUNCTION

FUNCTION dunw (X) 'Reduces big angles to <360°
dunw = (X - (INT(X / 360) * 360))
END FUNCTION
'_____
```

# Appendix 3 - Bibliography

| | | |
|---|---|---|
| A Guide to Small Boat Radio | M. Harris | Adlard Coles (UK) |
| All About Marine Compasses and their Adjustment | A. Pickles | Nadar Pty Ltd., PO Box 320 Freemantle, Western Australia, 6160 |
| Astro Navigation by Pocket Computer | M. Harris | Adlard Coles (UK) |
| The Deviant Compass | Capt. Mick Putney | The Prudent Mariner Inc |
| Astronomy with your Personal Computer | P. Duffett-Smith | Cambridge University Press. |
| Compact Data for Navigation and Astronomy | B. D. Yallop & C. Y. Hohenkerk | H.M.S.O |
| Compass Adjusting for Small Craft. What To Do How To Do It. | S. Kaufman | Surfside Harbor Associates |
| Compass Work | J. F. Kemp & P. Young | Heineman Newnes |
| Handbook of Magnetic Compass Adjustment | Publication No 226 | US Defense Mapping Agency |
| Magnetism From Lodestone to Polar Wandering. | D. S. Parasnis | Hutchinson 1961 |
| Oceanography and Seamanship | William G. Van Dorn | Dod, Mead & Co New York 1974 |
| Piloting & Dead Reckoning | H. H. Shufeldt & G. D. Dunlap | Naval Institute Press |
| Understanding Weatherfax | M. Harris | Adlard Coles (UK) |

## Magnetic Data Chart References

**United States Defense Mapping Agency**

| Stock Number | Title |
|---|---|
| 30 | The Magnetic Inclination or Dip, Epoch 1985.0, North & South Polar areas |
| 33 | The Horizontal intensity of the Earth's Magnetic Field, Epoch 1985.0, North & South Polar areas. |
| 36 | The Vertical Intensity of the Earth's Magnetic Field, Epoch 1985.0 |
| 39 | The Intensity of the Earth's Magnetic Field, Epoch 1985.0 |
| 42 | Magnetic Variation Chart of the World, Epoch 1985.0 |
| 43 | Magnetic Variation, epoch 1985.0, North & South Polar Areas. |

**United States Defense Mapping Agency**

| Chart Number | Title |
|---|---|
| 5374 | Magnetic Variation, 1995, and Annual rates of change. The World. |
| 5375 | Magnetic Variation, 1995, and Annual rates of change. North Atlantic Ocean and Mediterranean sea. |
| 5376 | Magnetic Variation, 1995, and Annual rates of change. South Atlantic Ocean. |
| 5377 | Magnetic Variation, 1995, and Annual rates of change. North Pacific Ocean. |
| 5378 | The World. Vertical Magnetic Intensity 1985 and annual rates of change. |
| 5379 | The World. Horizontal Magnetic Intensity 1985 and annual rates of change. |
| 5380 | The World. North Magnetic Intensity 1985 and annual rates of change. |
| 5381 | The World. East Magnetic Intensity 1985 and annual rates of change. |
| 5382 | The World. Total Magnetic Intensity 1985 and annual rates of change. |
| 5383 | The World. Magnetic dip 1985 and annual rates of change. |
| 5384 | Magnetic Variation 1990 and annual rates of change. Polar Regions |
| 5385 | Magnetic Variation 1995 and annual rates of change. Indian Ocean |
| 5399 | Magnetic Variation 1995 and annual rates of change. South Pacific Ocean |

# Appendix 4 - Suppliers and Web Sites

**The Henry Dietz Co., Inc**
Manufacturers of compass adjustment tools
14-26 28th Avenue Long Island City, New York 11102-3692 USA
www.idt.net/~hgdietz/

**Frequently Asked Questions on Magnetic Declination - Chris M Goulet.**
A very comprehensive and thoughtful set of questions from an experienced compass user. Covers technical points on the Earth's magnetic field, compass use.
www.cam.org/~gouletc/decl_faq.html

**KVH Industries, Inc.**
110 Enterprise Center, Middletown, RI 02842 USA
www.kvh.com

**National Geophysical Data Centre**
A primary source of Geomagnetic data including the Geomag. The site includes extensive links to other related resources.
www.ngdc.noaa.gov

**Pangolin Communications**
Marine software (incl. C-Swing compass adjuster) and related material.
www.pangolin.co.nz

**Autonav Marine Systems**
Essay by Paul Wagner on the use of Flux gate compasses in autopilots
http://www.helix.net/marine/autonav/anessay.html

# Index

# Notes

# Notes

# Notes